D1092741

SYSTEM SIMULATION

GEOFFREY GORDON

IBM CORPORATION
NEW YORK SCIENTIFIC CENTER

PRENTICE-HALL, INC.
ENGLEWOOD CLIFFS, NEW JERSEY

13-881805-3

Library of Congress Catalog Card Number: 77-87262

Current printing (last digit): 10   9   8   7   6   5   4   3   2   1

Printed in the United States of America

Prentice-Hall
Series in Automatic Computation
George Forsythe, editor

SYSTEM
SIMULATION

PRENTICE-HALL INTERNATIONAL, INC., *London*
PRENTICE-HALL OF AUSTRALIA, PTY. LTD., *Sydney*
PRENTICE-HALL OF CANADA, LTD., *Toronto*
PRENTICE-HALL OF INDIA PRIVATE LTD., *New Delhi*
PRENTICE-HALL OF JAPAN, INC., *Tokyo*

PREFACE

The use of simulation is expanding rapidly, and an understanding of the technique is becoming increasingly important to persons working in many disciplines. At the same time, a variety of general-purpose simulation programming languages have become available. This book is designed to introduce the reader to the principles of simulation and the application of several simulation languages to system studies. Examples are drawn from many different fields; no specialized knowledge in any particular discipline is assumed. The reader, however, is assumed to have been introduced to computer programming and, in certain sections, to have some knowledge of FORTRAN. A familiarity with probability and statistics will also be helpful, although two chapters review the required theory.

Chapters 1 and 2 introduce the principles of model-building and the technique of system simulation. The following three chapters are devoted to the simulation of continuous systems. Chapter 3 introduces a digital-analog simulation language, 1130/CSMP, and a continuous system simulation language, 360/CSMP. Simple examples of applications in engineering and biological systems are given. The topic of industrial dynamics, which is concerned primarily with the application of continuous system simulation to industrial and economic systems, is discussed in chapters 4 and 5. The first of these two chapters uses 360/CSMP to program examples, and the second discusses DYNAMO, a language specifically designed for industrial dynamics.

The following five chapters are devoted to the techniques of modeling and simulating discrete systems. Chapters 6 and 7 first review the probability and statistics theory needed to construct discrete system models. A hand-computed simulation of a simple telephone system is then given in chapter 8. The chapter concludes with the derivation of a flow-chart from which the problem can be programmed, together with a discussion of the tasks involved in the programming. Chapter 9 describes in detail a FORTRAN program written from the flow-chart derived in chapter 8. The following chapter expands the simulation and, in so doing, introduces some of the programming techniques used in writing simulation programs.

Two simulation languages designed principally for discrete systems are then discussed in the next four chapters. An introduction to GPSS is given in chapter 11, using a series of simple factory models as examples. Further examples are given in chapter 12, including the telephone system previously programmed in FORTRAN. The same approach is taken in describing SIMSCRIPT in chapters 13 and 14. The first introduces the language, again using the telephone system as an example, and the second expands the description.

The last chapter concerns the statistical problems of verifying simulation results and describes several methods applied to their solution. The positioning of this chapter at the end of the book is arbitrary; it could be studied at any point following chapter 7.

This book was designed principally for a one-semester graduate course. However, two or three chapters will need to be omitted to fit a course of that size. For this reason, the portions of the book concerned with industrial dynamics, FORTRAN, GPSS, and SIMSCRIPT have each been written as two chapters. In each case the first chapter is self-contained, giving the option of dropping some of the second chapters according to the students' interests and background. The full text, supplemented by term projects, should be sufficient for a two-semester course. All the programming information needed for the worked examples and the exercises has been included. It has not, however, been feasible to describe any of the languages in full detail. Readers planning to use a language extensively should supplement the text with the full programming information, for which references are given.

All the languages which are described are readily available, and that has been a main consideration in their choice. However, it is unlikely that all six languages are available at one place. In the case of the continuous simulation languages, only the basic elements of the programs have been used. It should not be difficult to convert the sections concerned with continuous systems and industrial dynamics to any other continuous system language that may be more readily available. It is not so easy to convert from one discrete language to another, but GPSS and SIMSCRIPT are probably the two most widely used simulation languages.

The selection of exercises that do not overtax the programming skill of the student presents difficulties. The approach that has been taken is to use exercises that require the text examples to be modified or expanded. They do not involve much programming, but they require an understanding of both the language and the model structure.

I would like to thank Mssrs. R. R. Coveyeau and D. R. Irving of the Oak Ridge National Laboratory for their assistance in discussing random number generators. In addition to the patient support that an author gratefully receives from his wife, my wife also gave me invaluable technical assistance in preparing the text. I would also like to thank Mr. N. J. Staniecki for his help in preparing the manuscript.

G.G.

CONTENTS

6

PROBABILITY CONCEPTS IN SIMULATION

7

ARRIVAL PATTERNS AND SERVICE TIMES

8

DISCRETE SYSTEM SIMULATION

9

SIMULATION WITH FORTRAN

10

SIMULATION PROGRAMMING TECHNIQUES

11

INTRODUCTION TO GPSS

12

GPSS EXAMPLES

13

INTRODUCTION TO SIMSCRIPT

14

MANAGEMENT OF SETS IN SIMSCRIPT

15

VERIFICATION OF SIMULATION RESULTS

Contents

1

SYSTEM MODELS

The term *system* is used in such a wide variety of ways that it is difficult to produce a definition broad enough to cover the many uses and, at the same time, concise enough to serve a useful purpose. We begin, therefore, with a simple definition of a system and expand upon it by introducing some of the terms that are commonly used when discussing systems. A *system* is defined as an aggregation or assemblage of objects joined in some regular interaction or interdependence. While this definition is broad enough to include static systems, the principal interest will be in dynamic systems where the interactions cause changes over time.

As an example of a conceptually simple system, consider an aircraft flying under the control of an autopilot (see Fig. 1-1). A gyroscope in the autopilot detects the difference between the actual heading and the desired heading. It sends a signal to move the control surfaces. In response to the control surface movement, the airframe steers towards the desired heading.

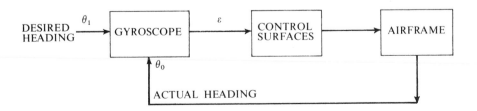

Figure 1-1. An aircraft under autopilot control.

As a second example, consider a factory that makes and assembles parts into a product (see Fig. 1-2). Two major components of the system are the fabrication department making the parts and the assembly department producing the products. A purchasing department maintains a supply of raw materials and a shipping department dispatches the finished products. A production control department receives orders and assigns work to the other departments.

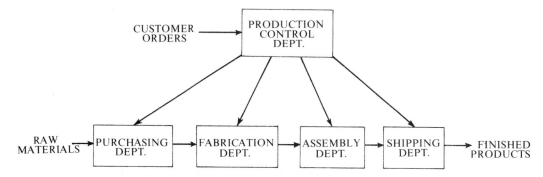

Figure 1-2. A factory system.

In looking at these systems, we see that there are certain distinct objects, each of which possesses properties of interest. There are also certain interactions occurring in the system that cause changes in the system. The term *entity* will be used to denote an object of interest in a system; the term *attribute* denotes a property of an entity. There can, of course, be many attributes to a given entity. Any process that causes changes in the system will be called an *activity*. The term *state of the system* will be used to mean a description of all the entities, attributes, and activities as they exist at one point in time. The progress of the system is studied by following the changes in the state of the system.

In the description of the aircraft system, the entities of the system are the airframe, the control surfaces, and the gyroscope. Their attributes are such factors as speed, control surface angle, and gyroscope setting. The activities are the driving of the control surfaces and the response of the airframe to the control surface movements. In the factory system, the entities are the departments, orders, parts, and products. The activities are the manufacturing processes of the departments. Attributes are such factors as the quantities for each order, type of part, or number of machines in a department.

Figure 1-3 lists examples of what might be considered entities, attributes, and activities for a number of other systems. If we consider the movement of cars as a traffic system, the individual cars are regarded as entities, each having as attributes its speed and distance traveled. Among the activities is the driving of a car. In the case of a bank system, the customers of the bank are entities with the balance of their account and their credit status as attributes. A typical activity would be the action of making a deposit. Other examples are shown in Fig. 1-3.

The figure does not show a complete list of all entities, attributes, and activities for the systems. In fact, a complete list cannot be made without knowing the purpose of the system description. Depending upon that purpose, various aspects of the system will be of interest and will determine what needs to be identified.

SYSTEM	ENTITIES	ATTRIBUTES	ACTIVITIES
TRAFFIC	CARS	SPEED DISTANCE	DRIVING
BANK	CUSTOMERS	BALANCE CREDIT STATUS	DEPOSITING
COMMUNICATIONS	MESSAGES	LENGTH PRIORITY	TRANSMITTING
SUPERMARKET	CUSTOMERS	SHOPPING LIST	CHECKING-OUT

Figure 1-3. Examples of systems.

1-2
System Environment

A system is often affected by changes occurring outside the system. Some system activities may also produce changes that do not react on the system. Such changes occurring outside the system are said to occur in the *system environment*. An important step in modeling systems is to decide upon the boundary between the system and its environment. The decision may depend upon the purpose of the study.

In the case of the factory system, for example, the factors controlling the arrival of orders may be considered to be outside the influence of the factory and therefore part of the environment. However, if the effect of supply on demand is to be considered there will be a relationship between factory output and arrival of orders, and this relationship must be considered an activity of the system. Similarly, in the case of a bank system, there may be a limit on the maximum interest rate that can be paid. For the study of a single bank, this would be regarded as a constraint imposed by the environment. In a study of the effects of monetary laws on the banking industry, however, the setting of the limit would be an activity of the system.

The term *endogenous* is used to describe activities occurring within the system and the term *exogenous* is used to describe activities in the environment that affect the system. A system for which there is no exogenous activity is said to be a *closed* system in contrast to an *open* system which does have exogenous activities.

One other distinction that needs to be drawn between activities depends upon the manner in which they can be described. Where the outcome of an activity can be described completely in terms of its input, the activity is said to be *deterministic*. Where the effects of the activity vary randomly over various outcomes, the activity is said to be *stochastic*.

The randomness of a stochastic activity would seem to imply that the activity is part of the system environment since the exact outcome at any time is not known. However, the randomness can often be measured and stated in the form of a probability distribution. If the occurrence of such an activity is under control of the system, it will be regarded as endogenous. If, however, the *occurrence* of the activity is random, it will constitute part of the environment. For example, in the case of the factory, the time taken for a machining operation may need to be described by a probability distribution but machining would be considered to be an endogenous activity. On the other hand, there may be power failures at random intervals of time. They would be the result of an exogenous activity.

1-3

Continuous and Discrete Systems

The aircraft and factory systems used as examples in Sec. 1-1 respond to environmental changes in different ways. The movement of the aircraft occurs smoothly, whereas the changes in the factory occur discontinuously. The ordering of raw materials or the completion of a product, for example, occurs at specific points in time.

Systems such as the aircraft, in which the changes are predominantly smooth, are called *continuous systems*. Systems like the factory, in which changes are predominantly discontinuous, will be called *discrete systems*. Few systems are wholly continuous or discrete. The aircraft, for example, may make discrete adjustments to its trim as altitude changes, while, in the factory example, machining proceeds continuously, even though the start and finish of a job are discrete changes. However, in most systems one type of change predominates, so that systems can usually be classified as being continuous or discrete.

Generally, a description of a continuous system will be in the form of continuous equations showing how the system attributes change with time. A description of a discrete system is concerned with the events producing changes in the state of the system. However, the type of description does not necessarily coincide with the type of system. The study of continuous systems will sometimes be simplified by considering the changes to occur as a series of discrete steps. Models of economic systems, for example, do not usually follow the flow of money and goods continuously; they consider the changes over regular intervals of time. In addition, the description of discrete systems is often simplified by considering the changes to occur continuously. The output of the factory, for example, may be described as a continuous variable, ignoring the discrete changes that occur as products are finished. The description of the system, therefore, is more important than the actual nature of the system.

To study a system, it is, of course, possible to experiment with the system itself. The objective of many system studies, however, is to predict how a system will perform before it is built. Clearly, it is not feasible to experiment with a system while it is in this hypothetical form. An alternative that is sometimes used is to construct a number of prototypes and test them, but this can be very expensive and time consuming. Even with an existing system, it is likely to be impossible or impractical to experiment with the actual system. For example, it is not feasible to study economic systems by arbitrarily changing the supply and demand of goods. Consequently, system studies are generally conducted with a model of the system. For the purpose of most studies, it is not necessary to consider all the details of a system; so a model is not only a substitute for a system, it is also a simplification of the system.

We define a *model* as the body of information about a system gathered for the purpose of studying the system. Since the purpose of the study will determine the nature of the information that is gathered, there is no unique model of a system. Different models of the same system will be produced by different analysts interested in different aspects of the system or by the same analyst as his understanding of the system changes.

The task of deriving a model of a system may be divided broadly into two subtasks: establishing the model structure and supplying the data. Establishing the structure determines the system boundary and identifies the entities, attributes, and activities of the system. The data provides the values the attributes can have and defines the relationships involved in the activities. The two jobs of creating a structure and providing the data are defined as parts of one task rather than as two separate tasks, because they are usually so intimately related that neither can be done without the other. Assumptions about the system direct the gathering of data and analysis of the data confirms or refutes the assumptions. Quite often, the data gathered will disclose an unsuspected relationship that changes the model structure.

To illustrate this process, consider the following description of a supermarket.

Shoppers needing *several items* of shopping *arrive* at a supermarket. They *get* a *basket*, if one is *available*, carry out their *shopping*, and then *queue* to *check-out* at one of the *several counters*.

After checking-out, they *return* the *basket* and *leave*.

Certain words have been italicized because they are considered to be key words that point out some feature of the system that must be reflected in the model. Essentially the same description is rewritten in Fig. 1-4 to identify the entities, attributes, and activities. Notice that the concept of a supermarket as a whole does not appear as an entity. It defines the system boundary and therefore distinguishes between the system and its environment. The arrival of customers in this description of the system will be regarded as an exogenous activity affecting the system from the environment. If, in contrast, the study objectives include analyzing the effects of car parking facilities on supermarket business, the boundary of the system would need to include the parking lot. The arrival of a customer in the supermarket depends upon finding a parking space, which can depend upon the departure of customers. Customer arrivals in the supermarket then become an endogenous activity; the arrival of cars becomes an exogenous activity.

ENTITY	ATTRIBUTE	ACTIVITY
SHOPPER	NO. OF ITEMS	
		ARRIVE
		GET
BASKET	AVAILABILITY	
		SHOP
		QUEUE
		CHECK-OUT
COUNTER	NUMBER OCCUPANCY	
		RETURN
		LEAVE

Figure 1-4. Elements of a supermarket model.

Other decisions about the system study objectives are implied in the model. The number of items of shopping is represented as an attribute of the shopper but no distinction has been made about the type of item. Secondly, no provision has been made in the system model for the effects of congestion on shopping time. If these decisions are not in keeping with the study objectives, another form of model must be used. In the first case, where type of item is to be distinguished, it is necessary to define several attributes for each customer, one for each type of item to be purchased. In the second case, where allowance for congestion must be made, two approaches could be taken. It may be necessary to introduce new entities representing the various sections of the supermarket and establish as attributes the number of customers they can serve simultaneously. Alternatively, the activity of shopping could be represented by a function in which shopping time depends upon the number of shoppers in the supermarket.

It is not suggested that Fig. 1-4 represents a formal process by which a transition can be made from a verbal description of a system to the structure of a model. It merely illustrates the process involved in forming a model.

Many types of models have been used in system studies and have been classified in a number of ways. The classification is sometimes made in terms of the nature of the system they model, such as continuous versus discrete or deterministic versus stochastic. For the purpose of this text, models will be treated as *physical models* or *mathematical models*.

A second distinction will be between *static models* and *dynamic models*. In the case of mathematical models, a third distinction is the technique employed in solving the model. A distinction is made between *analytic* and *numeric* methods. The classifications of models that have been described are illustrated in Fig. 1-5. As will be discussed in the next chapter, system simulation

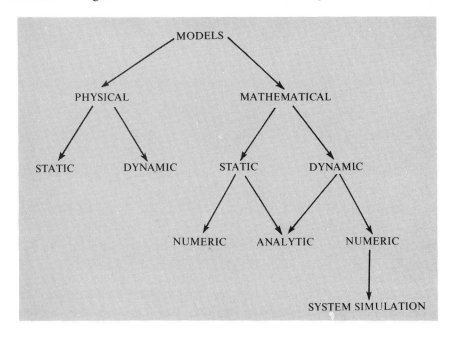

is considered to be a numeric technique using dynamic mathematical models, so system simulation is shown under the heading of numeric computation. For the sake of completeness, the following two sections will be devoted to brief discussions of the other types of model.

1-6

Physical Models

When discussing system models in the previous sections, it was not intended to imply that a model is necessarily a mathematical description of a system. It is possible to construct a physical model whose behavior represents the system being studied. The attributes of the system entities are represented by such physical measures as a voltage or the position of a shaft. The activities of the system are reflected in the physical laws that drive the model.

The best known examples of physical models are scale models used in wind tunnels and water tanks to study the design of aircraft and ships. Well-established laws of similitude allow accurate deductions about the performance of a full-sized system to be made from the scale model. Other types of physical models have been described as iconic models, that is, models which "look like" the system they model; for example, the models of molecular structures that are made from spheres representing atoms with rods representing atomic bonds. Both scale models and iconic models are examples of static physical models.

Dynamic physical models rely upon an analogy between the system being studied and some other system of a different nature, the analogy usually depending upon an underlying similarity in the forces governing the behavior of the systems. To illustrate this type of physical model, consider the two systems shown in Fig. 1-6. Figure 1-6(a) represents a mass that is subject to an

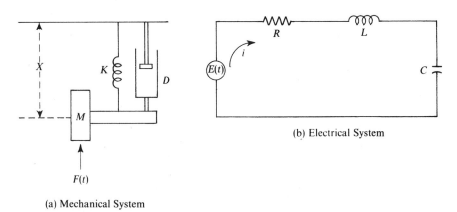

(b) Electrical System

(a) Mechanical System

Figure 1-6. Analogy between mechanical and electrical systems.

applied force $F(t)$ varying with time, a spring whose force is proportional to its extension or contraction, and a shock absorber that exerts a damping force proportional to the velocity of the mass. The system might, for example, represent the suspension of an automobile wheel when the automobile body is assumed to be immobile in a vertical direction. It can be shown that the motion of the system is described by the following differential equation:

$$M\ddot{x} + D\dot{x} + Kx = KF(t)$$

where x is the distance moved,
  M is the mass,
  K is the stiffness of the spring,
  D is the damping factor of the shock absorber.

Figure 1-6(b) represents an electrical circuit with an inductance L, a resistance R, and a capacitance C, connected in series with a voltage source that varies in time according to the function $E(t)$. If q is the charge on the capacitance, it can be shown that the behavior of the circuit is governed by the following differential equation:

$$L\ddot{q} + R\dot{q} + \frac{q}{C} = \frac{E(t)}{C}$$

Inspection of these two equations shows that they have exactly the same form and that the following equivalences occur between the quantities in the two systems:

Displacement	x	Charge	q
Velocity	$\dot{x}$	Current	$I\,(=\dot{q})$
Force	F	Voltage	E
Mass	M	Inductance	L
Damping factor	D	Resistance	R
Spring stiffness	K	1/Capacitance	$1/C$

The mechanical system and the electrical system are models of each other, and the performance of either can be studied with the other. In practice, it is simpler to modify the electrical circuit than to change the mechanical system, so it is more likely that the electrical system will have been built to study the mechanical system. If, for example, a car wheel is considered to bounce too much with a particular suspension system, the electrical model will demonstrate this fact by showing that the charge (and, therefore, the voltage) on the condenser oscillates excessively. To predict what effect a change in the shock absorber or spring will have on the performance of the car, it is only necessary to change the values of the resistance or condenser in the electrical circuit and observe the effect on the way the voltage varies.

If, in fact, the mechanical system were as simple as illustrated, it could be studied by solving the mathematical equation derived in establishing the analogy. However, effects can easily be introduced that would make the mathematical equation difficult to solve. For example, if the motion of the wheel is limited by physical stops, a non-linear equation that is difficult to solve will be needed to describe the system. It is easy to model the effect electrically by placing limits on the voltage that can exist on the capacitance. Even where the mathematical equations describing a system can be solved, the scale of the system might make it more convenient to take measurements on a physical system rather than carry out the necessary computations. Vibrations in structures are often studied in this way, even though it may be possible to solve the model equations.

The analogy made between the mechanical and electrical systems was established by showing that both systems obey the same mathematical equation. In practice, once the analogy has been established between resistance, inductance, and capacitance on the one hand, and damping, mass, and spring stiffness on the other, models are built directly from electrical components without resorting to mathematical equations. Examples of electrical currents being used in the analysis of mechanical systems will be found in Ref. (1). Other examples are the study of heat exchange problems (2) and the human nervous system (3).

1-7

Mathematical Models

In a *mathematical model*, the entities of a system and their attributes are represented by mathematical variables. The activities are described by mathematical functions that inter-relate the variables. Mathematical models are considered to be either static or dynamic.

A static model displays the relationships between the system attributes when the system is in equilibrium. If the point of equilibrium is changed by altering one or more of the attributes, the model enables the new values for all the attributes to be derived but does not show the way in which they changed to their new values.

The model may be solved analytically or it may be necessary to solve it numerically, depending upon the nature of the model.

As an example, consider the following simple mathematical model of the national economy (4). Let,

C be consumption,
I be investment,
T be taxes,
G be government expenditure,
Y be national income.

$$C = 20 + 0.7(Y - T)$$

$$I = 2 + 0.1Y$$

$$T = 0.2Y$$

$$Y = C + I + G$$

All quantities are expressed in billions of dollars.

Since there are four equations in five variables, if any one quantity is given, the equations can be solved analytically to derive the value of the other quantities. On the other hand, if the problem is to find the levels of government expenditure G and taxes T that maximize national income Y, the problem is one that needs to be solved numerically.

A dynamic mathematical model allows the changes of system attributes to be derived as a function of time. The derivation may be made with an analytical solution or with a numerical computation, depending upon the complexity of the model. The equation that was derived to describe the behavior of a car wheel is an example of a dynamic mathematical model; in this case, an equation that can be solved analytically. It is customary to write the equation in the form

$$\ddot{x} + 2\zeta\omega\dot{x} + \omega^2 = \omega^2 F(t)$$

where $2\zeta\omega = D/M$ and $\omega^2 = K/M$.

Expressed in this form, solutions can be given in terms of the variable ωt. Figure 1-7 shows how x varies in response to a steady force applied at time $t = 0$ as would occur, for instance, if a load were suddenly placed on the car. Solutions are shown for several values of ζ, and it can be seen that when ζ is less than 1, the motion is oscillatory.

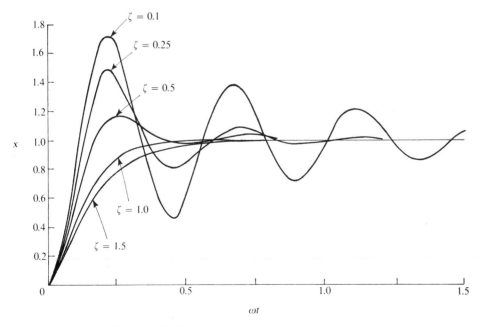

Figure 1-7. Solutions of second order equations.

The factor ζ is called the *damping ratio* and, when the motion is oscillatory, the frequency of oscillation is determined from the formula

$$\omega = 2\pi f$$

where f is the number of cycles per second.

Suppose a case is selected as representing a satisfactory frequency and damping. The relationships given above between ζ, ω, M, K, and D show how to select the spring and shock absorber to get that type of motion. For example, the condition for the motion to occur without oscillation requires that $\zeta \geqslant 1$. It can be deduced from the definition of ζ and ω that the condition requires that $D^2 \geqslant 4MK$.

Dynamic mathematical models that can be solved analytically and give practical results are not very common. More often, the model will have to be solved by numeric methods and, as indicated by Fig. 1-5, simulation is one such method. As an example of a non-simulation method of solving such models, consider the problem of determining the conditions under which an airframe will vibrate. A dynamic model can be established to describe the response of the airframe to aerodynamic forces and, in general, it cannot be solved analytically. The question of whether there will be vibrations or not can be determined numerically by evaluating the roots of certain associated equations. The numerical calculations will not constitute simulation, as it will be defined, since the calculation does not follow the motion of the airframe over time. It is simply a procedure for determining the roots of an equation.

1-8

Principles Used in Modeling

It is not possible to provide rules by which mathematical models are built, but a number of guiding principles can be stated. They do not describe distinct steps carried out in building a model. They describe different viewpoints from which to judge the information to be included in the model.

(a) Block-building

The description of the system should be organized in a series of blocks, or subsystems. The aim in constructing the blocks is to simplify the specification of the interactions within the system. Each block describes a part of the system that depends upon a few, preferably one, input variables and results in a few output variables. The system as a whole can then be described in terms of the interconnections between the blocks. Correspondingly, the system can be represented graphically as a simple block diagram.

The description of a factory given in Fig. 1-2 is a typical example of a block diagram. Each department of the factory has been treated as a separate block, with the inputs and outputs being the work passed from department to department. The fact that the departments might occupy the same floor space and might use the same personnel or the same machines has been ignored.

(b) Relevance

The model should only include those aspects of the system that are relevant to the study objectives. As an example, if the factory system study aims to compare the effects of different operating rules on efficiency, it is not relevant to consider the hiring of employees as an activity. While irrelevant information in the model may not do any harm, it should be excluded because it increases the complexity of the model and causes more work in solving the model.

(c) Accuracy

The accuracy of the information gathered for the model should be considered. In the aircraft system, for example, the accuracy with which the movement of the aircraft is described depends upon the representation of the airframe. It may suffice to regard the airframe as a rigid body and derive a very simple relationship between control surface movement and aircraft heading, or it may be necessary to recognize the flexibility of the airframe and make allowance for vibrations in the structure. An engineer responsible for estimating the fuel consumption may be satisfied with the simple representation. Another engineer, responsible for considering the comfort of the passengers, needs to consider vibrations and will want the detailed description of the airframe.

(d) Aggregation

A further factor to be considered is the extent to which the number of individual entities can be grouped together into larger entities. The general manager of the factory may be satisfied with the description that has been given. The production control manager, however, will want to consider the shops of the departments as individual entities.

In some studies, it may be necessary to construct artificial entities through the process of aggregation. For example, an economic or social study will usually treat a population as a number of social classes and conduct a study as though each social class were a distinct entity.

Similar considerations of aggregation should be given to the representation of activities. For example, in studying a missile defense system, it may not be necessary to include the details of computing a missile trajectory for each firing. It may be sufficient to represent the outcome of many firings by a probability function.

Exercises   1-1   Extract from the following description the entities, attributes, and activities of the system. Ships arrive at a port. They dock at a berth if one is available; otherwise, they wait until one becomes available. They are unloaded by one of several work gangs whose size depends upon the ship's tonnage. A warehouse contains a new cargo for the ship. The ship is loaded and then departs. Suggest two exogenous events (other than arrivals) that may need to be taken into account.

1-2   Name three or four of the principal entities, attributes, and activities to be considered if you were to simulate the operation of (a) a gasoline filling station, (b) a cafeteria, (c) a barber shop.

1-3   A new bus route is to be added to a city, and the traffic manager is to determine how many extra buses will be needed. What are the three key attributes of the passengers and buses that he should consider? If the company manager wants to assess the effect of the new route on the transit system as a whole, how would you suggest he aggregate the features of the new line to form part of a total system model? Would you suggest a continuous or discrete model for the traffic manager and the general manager?

1-4   In the automobile wheel suspension system, it is found that the shock absorber damping force is not strictly proportional to the velocity of the wheel. There is an additional force component equal to D_2 times the acceleration of the wheel. Find the new conditions for ensuring that the wheel does not oscillate.

1-5   A woman does her shopping on Mondays, Wednesdays, and Fridays. If it is fine, she walks to the stores; otherwise, she takes a bus. She always takes the bus home. On Tuesdays, she visits her daughter, traveling there and back by bus. Assuming that information is available about the day of the week and the state of the weather, draw a flow chart of her movements.

1-6 In the aircraft system, suppose the control surface angle y is made to be A times the error signal. The response of the aircraft to the control surface is found to be $I\ddot{\theta}_0 + D\dot{\theta}_0 = Ky$. Find the conditions under which the aircraft motion is oscillatory.

1-7 Suppose the automobile body in the suspension system example is not stationary. Consider the body to have a mass of M_1, and assume that its motion is determined by the force of gravity and the reaction with the suspension system. Construct a model for the motions of the wheel and body.

Bibliography

1 Keropyan, K. K. (ed.), *Electrical Analogues of Pin-Jointed Systems*, New York: The Macmillan Company, 1965.

2 Barabaschi, S., M. Conti, L. Gentilini, and A. Mathis, "Heat Exchange Simulator," *Automatica*, II, No. 1 (June, 1964), 1–13.

3 Brain, A. E., "The Simulation of Neural Elements by Electrical Networks Based on Multi-Apertured Magnetic Cores," *Proc. IRE*, XL (Jan., 1961), 49–52.

4 Suits, Daniel B., "Forecasting and Analysis With an Econometric Model," *American Economic Review*, LII (1962), 104–132.

2

SYSTEM SIMULATION

Given a mathematical model of a system, it is sometimes possible to derive information about the system by analytic means, as was demonstrated in Sec. 1-7. Where this is not possible, it is necessary to use numerical computation methods for solving the equations. A rich variety of numerical computation methods have been developed for solving the equations of mathematical models. In the case of dynamic mathematical models, a particular technique that has come to be identified as system simulation is one in which all equations of the model are solved simultaneously with steadily increasing values of time.

We therefore define *system simulation* as the technique of solving problems by following the changes over time of a dynamic model of a system. The definition is broad enough to include the use of dynamic physical models, in which case the variables of the model are evaluated by physical measurements rather than numerical computations. It is not intended here to discuss further the use of physical models, so future references to simulation will be in terms of mathematical models.

Since the technique of simulation does not attempt to solve the equations of a model analytically, a mathematical model constructed for simulation purposes is usually of a different nature than one constructed for analytic techniques. When building a model for analytic solution, it is necessary to bear in mind the constraints set by the analytic technique and to avoid complicating the over-all model. Many gross assumptions may have to be made to meet these constraints.

A simulation model, however, can be constructed more freely. Typically, it is built in a series of sections corresponding to the block diagram method that was recommended in Sec. 1-8. Each section can be described mathematically in a straightforward and natural manner without undue concern for the complexity introduced by having many such sections. The equations, however, must be constructed and organized in a way that enables a routine procedure to be used for solving them simultaneously.

In continuous systems, where the prime interest is in smooth changes, sets of differential equations are generally used to describe the system. Simulations based on such models are said to be *continuous simulations*. Analog computers, which are described briefly in the next chapter, can solve sets of linear differential equations simultaneously and are used extensively for continuous simulation. Digital computers can perform the same function by using small time interval steps to integrate the equations.

For discrete systems, where the prime interest is in events, the equations are essentially logical equations stating the conditions for an event to occur. The simulation consists of following changes in the state of the system that result from the succession of events. Such simulations are said to be *discrete simulations*. It is possible to advance time in small increments and check at each step if any events are due to be executed. As a general rule, however, discrete simulation is carried out by deciding upon the sequence of events and advancing time to the next most imminent event.

2-2

Experimental Nature of Simulation

The simulation technique makes no specific attempt to isolate the relationships between any particular variables; instead, it observes the way in which all variables of the model change with time. Relationships between the variables must be derived from these observations. A simulation study of the automobile wheel suspension system that was analyzed in Sec. 1-7 would proceed by following the motion of the wheel under different conditions. The relationship between D, K, and M to prevent oscillation, which was previously discovered analytically, would have to be discovered by observing the values that result in the motion being non-oscillatory. Simulation is, therefore, essentially an experimental problem-solving technique. Many simulation runs have to be made to understand the relationships involved in the system, so the use of simulation in a study must be planned as a series of experiments.

The manner in which the simulation experiments proceed depends upon the nature of the study. Generally, system studies are of three main types: system analysis, system design, and what will be called system postulation.

Many studies, in fact, combine two or three of these aspects or alternate between them as the study proceeds. The term *system engineering* is frequently used to describe system studies where a combination of analysis and design is aimed at understanding, first, how an existing system works and then preparing system modifications to change the system behavior.

System analysis aims to understand how an existing system or a proposed system operates. The ideal situation would be that the investigator is able to experiment with the system itself. What is actually done is to construct a model of the system and, by simulation, investigate the performance of the model. The results obtained are interpreted in terms of system behavior.

In *system design* studies, the object is to produce a system that meets some specifications. Certain system components can be selected or planned by the designer, and, conceptually, he chooses a particular combination of components to construct a system. The proposed system is modeled and its performance predicted from knowledge of the model's behavior. If the predicted behavior compares favorably with the desired behavior, the design is accepted. Otherwise, the system is redesigned and the process repeated.

System postulation is characteristic of the way simulation is employed in social, economic, political, and medical studies, where the behavior of the system is known but the processes that produce the behavior are not. Hypotheses are made on a likely set of entities and activities that can explain the behavior. The study compares the response of the model based on these hypotheses with the known behavior. A reasonably good match naturally leads to the assumption that the structure of the model bears a resemblance to the actual system, and allows a system structure to be postulated. Very likely, the behavior of the model gives a better understanding of the system, possibly helping to formulate a refined set of hypotheses.

An example of this type of study designed to investigate the functioning of the liver in the human body is described in Ref. (1). When a chemical, thyroxine, is injected into the blood stream, it is carried to the liver. The liver can change thyroxine into iodine which is absorbed into the bile. However, neither the conversion nor the absorption occur instantaneously. Some of the thyroxine reenters the blood stream to be recirculated and returned to the liver. By using radioactive isotopes, it is possible to measure the rate at which thyroxine is removed from the blood stream, but the precise mechanism by which it is transferred from the blood to the liver and then to the bile was not known.

In the study, a mathematical model was constructed assuming that the body can be represented as three compartments and that the rates at which thyroxine is transferred between the compartments are proportional to the concentration of thyroxine in the compartments. Figure 2-1, reproduced from Ref. (1), illustrates the model and shows the assumed transfer coefficients between compartments. The compartments 1, 2, and 3 represent the blood vessels, the liver, and the bile, respectively. The model leads to three simple differential equations which are shown, together with their general solutions, in Fig. 2-1.

Figure 2-2, also reproduced from Ref. (1), shows the results of comparing actual measurements against the model predictions for one set of assumed values of the coefficients. It can be seen that the match between the results of

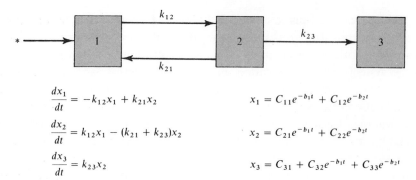

$$\frac{dx_1}{dt} = -k_{12}x_1 + k_{21}x_2 \qquad\qquad x_1 = C_{11}e^{-b_1t} + C_{12}e^{-b_2t}$$

$$\frac{dx_2}{dt} = k_{12}x_1 - (k_{21} + k_{23})x_2 \qquad\qquad x_2 = C_{21}e^{-b_1t} + C_{22}e^{-b_2t}$$

$$\frac{dx_3}{dt} = k_{23}x_2 \qquad\qquad x_3 = C_{31} + C_{32}e^{-b_1t} + C_{33}e^{-b_2t}$$

Figure 2-1. Mathematical model of the liver.

the theoretical model and the experiment is close, which suggests that the hypothesis on transfer rates is reasonably accurate and allows estimates of the transfer coefficients to be made. In fact, the particular study from which this example is quoted found a more comprehensive model fitted the experimental facts more closely.

The model used here was solved analytically and the study did not use simulation in the sense in which it has been defined. The example has been chosen to illustrate the approach used in system postulation. There are many examples of studies using the same general method where the mathematical model can only be solved by system simulation (2).

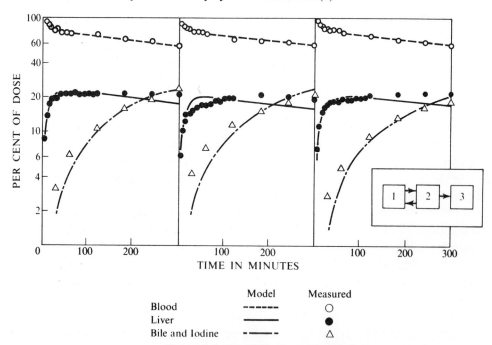

Figure 2-2. Results from model of the liver.

The application of simulation to many types of systems together with the different types of study result in many variations in the way a simulation study proceeds. Certain basic steps in the process can, however, be identified. The principal steps are considered to be:

1. Definition of the problem
2. Planning the study
3. Formulation of a mathematical model
4. Construction of a computer program for the model
5. Validation of the model
6. Design of experiments
7. Execution of simulation run and analysis of results.

The first two steps are to define the problem and plan the study. While these steps may seem to be obvious, they are none the less important. No study, simulation or otherwise, should proceed until a clear statement of the problem and the objectives of the study are established. Estimates can then be made of the work to be done and the time required. Nor is the usefulness of the plan finished when the study has started; the plan can control the progress of the work and prevent the study from becoming unbalanced by concentrating on one aspect of the problem at the expense of another. A common failing in simulation studies is becoming so engrossed in simulating that more detail is extracted from the simulation than is needed or can be supported by the data available.

The third step consists of constructing a model, a task that can be thought of as falling into two sub-tasks. The structure of the model must be established by deciding what aspects of the system performance are significant for the problem in hand, and data must be gathered to provide correct parameters for the model.

Given a mathematical model, the construction of a computer program for the model, the fourth step, is a relatively well-defined task. It is not necessarily an easy task and can be very time consuming, but the model establishes the specifications of what must be programmed. In practice, the question of how difficult it is to program a model often influences how the model is constructed. The tasks of producing a model and a computer program are likely to be carried out in parallel rather than in series.

The fifth step, validation of the model, is an area requiring a good deal of judgment. To a large extent, the problem is the complement of the formulation of the model. Inferences made in establishing the model are checked by observing if the model behaves as expected. Errors can, of course, occur when programming a model. Ideally, the errors of the model and programming errors should be separated by validating the mathematical model before embarking on the programming. This is not easy to do, however, because the reason for simulating in the first place is usually the fact that the mathematical model is intractable. It may be possible to solve special cases, for example, by removing all randomness, but as a rule the validation is carried out by examining the computer version of the model.

The sixth step is designing a set of experiments to meet the study objectives. A factor to be considered is the cost of running a computer model, since this may limit the number of runs that can be made. Even when this is not so, careful thought should be given to what runs are necessary. A common fault in simulation studies is that the system analyst becomes overwhelmed with a mass of computer output data gathered with no particular plan. The presence of random events in a simulation complicates the design of the experiments since the statistical significance of the results must be considered. This problem will be discussed in Chap. 15.

The final step in a system study is executing the simulation runs and interpreting the results. In a well-planned study a clear-cut set of questions will have been posed and the analysis aims to answer the questions.

2-4

Recurrence Models

The equations of a model become complex when each equation involves several variables of the problem simultaneously. One approach to simplifying the numerical calculations is to consider time to advance in uniform steps and organize the model in the form of a series of difference equations. Some of the equations are organized to show the current value of a variable in terms of the values of one or more variables in the previous interval. The variables described in the previous interval are called lagged variables. Properly organized, a set of equations in this form allows values of all variables in the present interval to be determined from the lagged variables. Time is then increased by one interval, and the computed values supply the lagged variables needed for the calculations of the next interval.

The method is particularly straightforward when the equations of the model are linear algebraic equations. This type of model is used widely in studies of economic systems and is called *econometric*. For example, the static mathematical model given in Sec. 1-7 can be made dynamic by selecting a fixed time interval, say one year, and expressing the equations in terms of lagged variables. The equations take the following form, where the suffix -1 is used to denote a lagged variable,

$$I = 2 + 0.1 Y_{-1}$$
$$T = 0.2 Y_{-1}$$
$$Y = C_{-1} + I_{-1} + G_{-1}$$
$$C = 20 + 0.7(Y_{-1} - T_{-1})$$

Given an initial set of values for all variables, the values of the variables at the end of one year can be derived. Taking these values as the new values of the lagged variables, the values can then be derived at the end of the second year, and so on. There are only four equations in five unknowns, so one variable must be specified for each interval.

It is not necessary, however, to lag all the variables as has just been done. Suppose there is one equation that expresses a single current variable in terms of lagged variables only. When this equation is solved, a second equation can be solved that involves the current value just derived from the first equation plus any lagged variables. A third equation can then involve the current values derived from the first two equations plus any lagged variables, and so on. Using this principle, a slight rearrangement of the model just given makes it depend upon only one lagged variable. Substituting for T and C in the original equation for Y gives

$$Y = 45.45 + 2.27(I + G)$$

The set of equations can then be written in the form:

$$I = 2 + 0.1Y_{-1}$$
$$Y = 45.45 + 2.27(I + G)$$
$$T = 0.2Y$$
$$C = 20 + 0.7(Y - T)$$

The only lagged variable is Y, and to solve the model an initial value of Y_{-1} must be given. With the lagged variable, the current value of I can be derived from the first equation. Suppose the values of G are supplied for all intervals. The next equation then gives the current value of Y from the current value of I. The next equation similarly gives the current value of T from the current value of Y, and the last equation gives the current value of C from current values of both Y and T. Taking the current value of Y as the new value of the lagged variable Y_{-1}, the calculations can then be repeated for the next interval of time.

Reflecting the manner in which the equations are solved, models of this type are called *recurrence models*. Constructing these models and, in particular, deciding which variables should be lagged requires careful analysis and a considerable literature exists on the subject. See, for example, Ref. (3). The models can become very large as more factors are included and, although this type of model can be solved by hand calculation, digital computers are used extensively to solve them.

2-5

Cobweb
Models

A particularly simple form of recurrence model occurs when only two equations are involved. Nevertheless, these simple models are of use in considering some of the gross effects involved in economic systems.

As an example, consider the following model for the marketing of a product. Let S be the supply, D be the demand, and P be the price of the product. The state of the system is considered at uniform steps of time, say 1 month intervals. The supply at time t is assumed to depend linearly upon the price obtained in the *previous* interval; the supply increasing with price. So that,

$$S = a + bP_{-1}, \qquad b > 0$$

The demand is assumed to depend upon *current* price and it decreases with increasing price.

$$D = c - dP, \qquad d > 0$$

For simplicity, assume the market is cleared each month, that is

$$D = S$$

This equation implies that the price changes to the level at which the supply equals the demand. The equations can be written in the form:

$$S = a + bP_{-1}$$
$$D = S$$
$$dP = c - D$$

They then meet the conditions laid down for a recurrence model and the successive states of the market can be calculated. Figure 2-3 shows the fluctuations of price for two cases as follows:

(a)	(b)
$P_0 = 1.0$	$P_0 = 5.0$
$a = 1.0$	$a = -2.4$
$b = 0.9$	$b = 1.2$
$c = 12.4$	$c = 10.0$
$d = 1.2$	$d = 0.9$

As can be seen, case (a) represents a stable market in which the market settles to a price of 5.43. Case (b) is unstable, with price fluctuating with increasing amplitude.

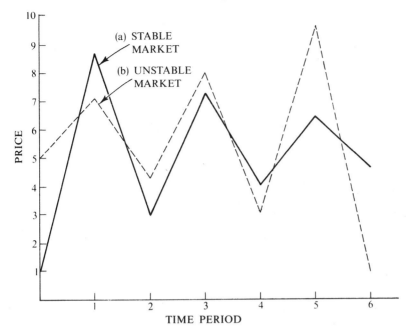

Figure 2-3. Fluctuations of market price.

Models of this type are called *cobweb models* (4) because of a particular way they can be solved graphically. The method is illustrated in Fig. 2-4 for the stable case just considered. The graph plots price on the horizontal axis and quantity on the vertical. Two straight lines plot the linear relationships between quantity and price; one expressing the demand as a function of price, the other giving supply as a function of price. Beginning with the initial price of 1.0, the corresponding supply is found on the supply line. This supply becomes the demand since the market is cleared, so a horizontal line is taken to the demand line to determine the price settled upon in the first period. That price determines the supply for the next period and the supply is derived by drawing a vertical line to the supply curve. The process repeats with each half turn around the spiral representing one period. The successive prices are marked as points 1, 2, ..., 6.

2-6

Simulation Programming Languages

The fact that simulation involves numerical computation naturally leads to the use of digital computers, and many programming languages have been designed for carrying out simulations. In general, these languages provide the user with a set of modeling concepts used to describe the system and a programming system that will convert the description into a computer program that performs the simulation. The user is thereby relieved of a substantial

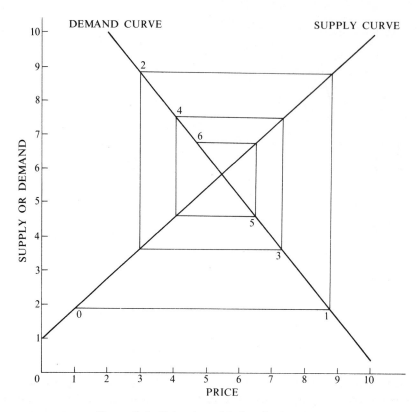

Figure 2-4. Cobweb model of market economy.

amount of detailed programming effort, particularly in the case of discrete simulation programs, where the task of managing the logical steps involved in executing events can become very involved. Reviews of the many languages that are available will be found in Refs. (5) and (6). Systems are classified as continuous or discrete, and the programming languages are usually designed for continuous or discrete systems. Later chapters will discuss three continuous system simulation languages, 1130/CSMP, 360/CSMP, and DYNAMO, and two discrete system simulation languages, GPSS and SIMSCRIPT.

2-7

System Simulation Applications

It is not feasible to describe in this text the many applications that have been made of simulation. It has been used in virtually all branches of science and engineering and descriptions of simulation studies will be found throughout the technical literature. Reference (7) contains an extensive bibliography of over 900 references to the theory and application of simulation. Some specific application areas that have been described in texts are as follows: business (8), economics (9), sociology (10), psychology (11), and politics (12) and (13).

2-1 Draw the cobweb model of Sec. 2-5 for the unstable conditions (case b). **Exercises**

2-2 Draw a cobweb model for the following market:

$$D = 12.4 - 1.2P$$

$$S = 8.0 - 0.6P_{-1}$$

$$P_0 = 1.0$$

2-3 Derive the mathematical conditions under which the market model of Sec. 2-5 is stable. Interpret the results graphically. (Hint: Express the equations in terms of the deviations of price and quantity from the equilibrium values).

2-4 The supplier in the market model of Sec. 2-5 learns to discount the swings in market price by adjusting the price upon which he bases the supply. He computes the supply from the given curve but uses the price $P_{t-1} - r(P_{t-1} - P_{t-2})$. Assume $P_0 = P_1 = 1.0$ and compute the market fluctuations. (The problem cannot be conveniently solved graphically. Use algebraic methods.)

2-5 Use a cobweb model to investigate a market in which the supply and demand functions are

$$D = \frac{17.91}{P^{1/2}} - 4.66$$

$$9S = 5.0(P_{-1} - 1)$$

Assume the market is always cleared.

2-6 Find the growth in national consumption for five years using the model given in Sec. 2-4. Assume the initial income Y_{-1} is 80 and take the government expenditure in the 5 years to be as follows:

Year	G
1	20
2	25
3	30
4	35
5	40

Bibliography   1   Hazelrig, Jane B., "The Impact of High-Speed Automated Computation on Mathematical Models," *Mayo Clinic Proc.*, XXXIX, No. 11 (Nov. 1964), 841–848.

2   Mesarovic, M. D. (ed.), *Systems Theory and Biology*, New York: Springer-Verlag, 1968.

3   Christ, Carl F., *Econometric Methods and Models*, New York: John Wiley and Sons, Inc., 1966.

4   Allen, R. G. D., *Macro-Economic Theory*, New York: St. Martin's Press, Inc., 1968.

5   Teichroew, Daniel, and John Francis Lubin, "Computer Simulation: Discussion of the Technique and Comparison of Languages," *Comm. ACM*, IX, No. 10 (Oct., 1966), 723–741.

6   Krasnow, H. S., and R. Merikallio, "The Past, Present and Future of General Simulation Languages," *Management Science*, XI, No. 2 (Nov., 1964), 236–267.

7   Bibliography on Simulation, IBM Corp., Form No. 320-0924, 1966.

8   Bonini, Charles P., *Simulation of Information and Decision Systems in the Firm*, Englewood Cliffs, N.J.: Prentice-Hall, Inc., 1963.

9   Orcutt, G. H., M. Greenberger, J. Korbel, and A. M. Rivlin, *Micro-Analysis of Socioeconomic Systems: A Simulation Study*, New York: Harper and Row, Publishers, 1961.

10   Borko, Harold (ed.), *Computer Applications in the Behavioral Sciences*, Englewood Cliffs, N.J.: Prentice-Hall, Inc., 1962.

11   Tompkins, Silvan, and Samuel Messick (eds.), *Computer Simulation of Personality*, New York: John Wiley and Sons, Inc., 1963.

12   Guetzkow, Harold, Chadwick F. Alger, Richard A. Brody, Robert C. Noel, and Richard C. Snyder, *Simulation in International Relations: Developments for Research and Teaching*, Englewood Cliffs, N.J.: Prentice-Hall, Inc., 1963.

13   Coplin, William D. (ed.), *Simulation in the Study of Politics*, Chicago, Ill.: Markham Publishing Co., 1968.

3

CONTINUOUS SYSTEM SIMULATION

3-1

Continuous System Models

A continuous system is one in which the predominant activities of the system cause smooth changes in the attributes of the system entities. When such a system is modeled mathematically, the variables of the model representing the attributes are controlled by continuous functions. The econometric model discussed in Sec. 2-4 is an example of a continuous model. That model described how the attributes of the system were related to each other in the form of linear algebraic equations. More generally, in continuous systems, the relationships describe the *rates* at which attributes change, so that the model consists of differential equations.

The simplest differential equation models have one or more linear differential equations with constant coefficients. It is then often possible to solve the model without the use of simulation, just as it is possible to solve many econometric models consisting of linear algebraic equations. Even so, the labor involved may be so great that it is preferable to use simulation techniques. However, when non-linearities are introduced into the model, it frequently becomes impossible or, at least, very difficult to solve the models. Simulation methods of solving the models do not change fundamentally when non-linearities occur. The methods of applying simulation to continuous models can therefore be developed by showing their application to models where the differential equations are linear and have constant coefficients, and then generalizing to more complex equations.

3-2

Linear   An example of a linear differential equation with constant coefficients was
Differential given in Sec. 1-6 to describe the wheel suspension system of an automobile.
Equations The equation derived was

$$M\ddot{x} + D\dot{x} + Kx = KF(t) \tag{3-1}$$

Note that the dependent variable x appears together with its first and second derivatives $\dot{x}$ and $\ddot{x}$, and that the terms involving these quantities are multiplied by constant coefficients and added. The quantity $F(t)$ is an input to the system depending upon the independent variable t. A linear differential equation with constant coefficients is always of this form, although derivatives of any order may enter the equation.

The simplest linear differential equation involves only the first derivative of a variable. Nevertheless, it is of considerable importance in modeling continuous systems since it represents the way in which many factors grow or decay. Consider, for example, the growth of a capital fund that is earning compound interest. If the growth rate is p (i.e., $100p\%$ interest), then the rate at which the fund grows is p times the current size of the fund. Expressed mathematically, where x is the current size of the fund,

$$\dot{x} = px$$

$$x = x_0 \quad \text{at } t = 0$$

This is a first-order differential equation whose solution is

$$x = x_0 e^{pt}$$

Curve A of Fig. 3-1 plots the response with p expressed as $1/T$. It can be seen that the fund grows indefinitely at an exponential rate. The same simple model represents the growth of a population, where the excess of birth rate over death rate is p. It can also represent a chemical reaction and many other phenomena.

A closely related model is one in which the magnitude of the variable is limited to some maximum value X, and the growth rate is proportional to the amount by which the variable falls short of the maximum. The equation for the model is

$$\dot{x} = p(X - x)$$

$$x = 0 \quad \text{at } t = 0$$

and the solution is

$$x = X(1 - e^{-pt})$$

The response is shown as curve B of Fig. 3-1. The model could represent the growth of a population where resources can only support a total population of X, or it could represent the growth of a company in a limited size market.

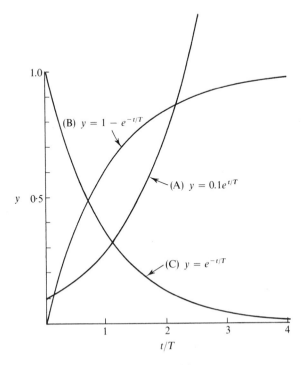

Figure 3-1. Exponential responses.

In other models, a variable decays from some initial value X at a rate proportional to the current value. The model is, then,

$$\dot{x} = -px$$
$$x = X \quad \text{at } t = 0$$

(3-2)

and the solution is

$$x = Xe^{-pt}$$

(3-3)

The response is shown as curve C of Fig. 3-1. The model represents, for example, the decay of nuclear material due to radiation.

3-3

As was pointed out in Chap. 1, many types of physical models based on analogies can be constructed. The specific example discussed there described

Analog Computers

the analogy between mechanical and electrical systems. In addition to analog models, there are many physical devices whose behavior is analogous to a mathematical operation such as addition or integration. Using these devices, it is possible to build analog computers to simulate systems. The computers employ a number of analog devices, each used to represent a mathematical operation on a particular variable. The technique of applying the computers consists of interconnecting the devices in a manner specified by a mathematical model of the system. Although analog computers are constructed from physical devices, it is more accurate to regard them as solving problems with mathematical models, rather than refer to them as physical models.

The most widely used form of analog computer is the electronic analog computer, based on the use of high gain dc (direct current) amplifiers (1). Voltages in the computer are equated to mathematical variables and the dc amplifiers can add and integrate the voltages. With appropriate circuits, an amplifier can be made to add several input voltages, each representing a variable of the model, to produce a voltage representing the sum of the input variables. Different scale factors can be used on the inputs to represent coefficients of the model equations. Such amplifiers are called *summers*. Another circuit arrangement produces an *integrator* for which the output is the integral with respect to time of a single input voltage or the sum of several input voltages. All voltages can be positive or negative to correspond to the sign of the variable represented. To satisfy the equations of the model, it is sometimes necessary to use a *sign inverter*, which is an amplifier designed to cause the output to reverse the sign of the input.

Electronic analog computers are limited in accuracy for several reasons. It is difficult to carry the accuracy of measuring a voltage beyond a certain point. Secondly, a number of assumptions are made in deriving the relationships for dc amplifiers, none of which is strictly true; so, amplifiers do not solve the mathematical model with complete accuracy. A particularly troublesome assumption is that there should be zero output for zero input. Another type of difficulty is presented by the fact that dc amplifiers have a limited dynamic range of output, so that scale factors must be introduced to keep within the range. As a consequence, it is difficult to maintain an accuracy better than 0.1 % in an electronic analog computer. Other forms of analog computers have similar problems and their accuracies are not significantly better.

A digital computer is not subject to the same type of inaccuracies. Virtually any degree of accuracy can be programmed and, with the use of floating-point representation of numbers, an extremely wide range of variations can be tolerated. Integration of variables is not a natural capability of a digital computer, as it is in an analog computer, so that integration must be carried out by numerical approximations. However, methods have been developed which can maintain a very high degree of accuracy.

A digital computer also has the advantage of being easily used for many different problems. An analog computer must usually be dedicated to one application at a time.

3-4

**Digital-
Analog
Simulators**

To avoid the disadvantages of analog computers, many programming systems, called *digital-analog simulators*, have been written. They allow a continuous system problem to be programmed on a digital computer in essentially the same way it is solved on an analog computer. A history and review of these programs is given in Ref. (2). They maintain the same general techniques developed to solve problems with analog computers, but in doing so they overcome the disadvantages of analog computers. Their advent has by no means replaced analog computers. The accuracy of analog computers is sufficient for many problems. Analog computers often provide a cheaper means of solving problems, particularly in the case of large-scale problems. They solve equations in a truly simultaneous manner while a digital computer must solve the equations serially. This often gives analog computers a considerable speed and cost advantage.

3-5

**The 1130
Continuous
System
Modeling
Program
(CSMP)**

The many digital-analog simulators that have been written are similar in their properties. They differ in programming details, in the range of features included beyond those basic for analog methods, and in the computer for which they are available. One particular program that will be described here is the Continuous System Modeling Program (CSMP). It is available for the IBM 1130 computer and referred to as 1130/CSMP, (3).

Problems are programmed as a block diagram, although they must eventually be reduced to a set of statements for entry into the computer. Usually, these statements are key-punched and loaded into the computer as an input deck but, with 1130/CSMP, as with several other digital-analog simulators, it is also possible to enter statements into the computer directly from a console.

There are altogether 25 block types, each represented by a symbol and coded in programming statements with a particular character. A list of 10 of the more commonly used 1130/CSMP blocks is shown in Fig. 3-2. Shown beside the symbols are equations describing the functions the blocks perform and the programming language symbols used to represent the blocks. The letter n included in the block symbols represents a block number to be provided by the user.

The block shown at the top of Fig. 3-2 is for a *weighted summer*. It corresponds to a dc amplifier summer except that it is limited to three inputs. The *summer* that follows differs from a dc amplifier summer in that no scale factors can be introduced but, unlike a dc amplifier, it is possible to specify that a

ELEMENT TYPE	LANGUAGE SYMBOL	DIAGRAMMATIC SYMBOL	DESCRIPTION
WEIGHTED SUMMER	W		$e_0 = P_1 e_1 + P_2 e_2 + P_3 e_3$
SUMMER	+		$e_0 = \pm e_1 \pm e_2 \pm e_3$ Only element where negative sign is permissible in configuration specification
INTEGRATOR	I		$e_0 = P_1 + \int(e_1 + e_2 P_2 + e_3 P_3)dt$
SIGN INVERTER	–		$e_0 = -e_i$
LIMITER	L		
NEGATIVE CLIPPER	N		
POSITIVE CLIPPER	P		
MULTIPLIER	X		$e_0 = e_1 e_2$
DIVIDER	/		$e_0 = e_1/e_2$
CONSTANT	K		$e_0 = P_1$

Figure 3-2. 1130/CSMP block types.

negative sign be attached to any of the three inputs. In effect, the summer is the same as a weighted summer with the restriction that the scale factors on the inputs be either $+1$ or -1. The *integrator* that follows can have three inputs, two of which can be scaled with a (positive) constant and one cannot. A third parameter is reserved for entering an initial value. Although a change of sign can be introduced at a summer, it is still convenient to have a *sign inverter* and that is provided by the next block type.

There follow in Fig. 3-2 three block types concerned with limiting the range a variable can take. The *limiter block* places an upper positive limit and a lower negative limit on a variable. The *negative clipper* and *positive clipper* prevent a variable from going negative or positive, respectively. Units capable of performing the same functions are generally available in an analog computer. Next are shown a *multiplier block* and a *divider block*. In the case of analog computers, these functions are difficult to supply since they require special equipment. Lastly is shown a *constant* block type for introducing a fixed value into the model.

1503282

Because digital-analog simulators (and analog computers) can add and integrate voltages they are particularly well suited to solving linear differential equations with constant coefficients. The general method used is to begin with the highest order derivative of the equation and produce the successively lower order derivatives by integration. Consider, for example, the second-order equation that has already been discussed:

$$M\ddot{x} + D\dot{x} + Kx = KF(t) \qquad (3\text{-}4)$$

Solving the equation for the highest order derivative gives

$$M\ddot{x} = KF(t) - D\dot{x} - Kx \qquad (3\text{-}5)$$

Suppose a variable representing the input $F(t)$ is supplied, and assume for the time being that there exist variables representing $-x$ and $-\dot{x}$. These three variables can be weighted and added with a weighted summer to produce a voltage representing $M\ddot{x}$ in accordance with Eq. (3-5). Integrating this variable with a scale factor of $1/M$ produces $\dot{x}$. Changing the sign produces $-\dot{x}$, which supplies one of the variables initially assumed; and a further integration produces $-x$, which was the other assumed variable. For convenience, a further sign inverter is included to produce $+x$ as an output. An 1130/CSMP block diagram to solve the problem in this manner is shown in Fig. 3-3. It should be pointed out that there are often several ways of drawing a block diagram, depending upon what outputs are of particular interest and which constants may need to be changed during a simulation study.

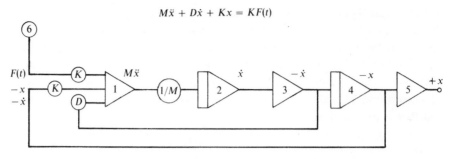

$$M\ddot{x} + D\dot{x} + Kx = KF(t)$$

Figure 3-3. Diagram for the automobile suspension problem.

Having constructed a block diagram, the preparation of an input deck for running a problem is made with the use of two coding forms. One form, the configuration data form, is used to prepare cards describing the blocks and their interconnections. The second form is for initial conditions and parameter data associated with the blocks. Figure 3-4 shows the coding for the automobile suspension problem.

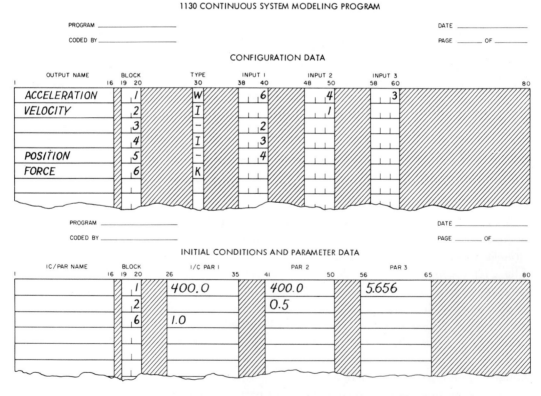

Figure 3-4. 1130/CSMP coding for the automobile suspension problem.

On the first form, there is one line for each block. A name may be given in columns 1–16 to the output of each block. The block number appears in columns 19 and 20 and the block type character appears in column 30. Three fields are used to record the numbers of the input blocks. Block numbers must be in the range 1–75, but they may be assigned arbitrarily and may be presented in any order on the coding form. All block numbers must be right-justified.

All parameters are initially set to zero, so unused parameter fields can be left blank. A name can be assigned to each set of parameters in columns 1–16. Columns 19 and 20 must carry the number of the block with which the parameters are associated, and three fields carry the values of the parameters. Parameters are entered as 1–6 decimal digit numbers plus a decimal point which must be present. Figure 3-4 shows the coding of the automobile suspension system for the case where

$$M = 2.0 \quad (1/M = 0.5)$$

$$K = 400.0$$

$$D = 5.656$$

One card is punched for each line of the forms. The parameter cards are placed behind the configuration deck, separated by a blank card, and the combined deck is placed behind some control cards also separated by a blank card. Another blank card is placed at the end of the parameter cards.

Any variable can become part of the output, and the user may tabulate values of the variable at repeated intervals. In addition, if the installed computer includes a plotter, the variables can be plotted against time. When the problem is loaded, the computer prints several messages at the console requesting the user to enter certain items of data. The information that is requested is:

(a) The time interval for integration
(b) The total (simulated) time for the run
(c) The block outputs to be tabulated
(d) The time intervals at which to print
(e) The block output to be plotted together with the maximum and minimum values to be established on the plotter

The automobile suspension problem was run with an integration interval of 0.005 seconds for a total time of 1.5 seconds. Printing was requested at intervals of 0.05 seconds. The curves previously shown as Fig. 1-7 were in fact the output from this problem. The curve marked $\zeta = 0.1$ is for the case coded in

Fig. 3-4. The other runs that are plotted were made by using the console to modify the value of D entered as parameter 1 of block 1. The values used were

ζ	D
0.25	14.14
0.5	28.28
1.0	56.56
1.25	70.70

3-7
Simultaneous Equations

When a model has more than one variable, it will consist of several equations. In effect, when such a problem is solved with a digital-analog simulator, a separate block diagram is drawn for each equation and, where necessary, interconnections are made between the block diagrams.

As an example, Fig. 3-5 shows an 1130/CSMP diagram for solving the model of the liver shown in Fig. 2-1. Three blocks each solve one of the three equations for X1, X2, and X3. Interconnections between the blocks introduce the value of X2 into the equation for X1 and vice versa.

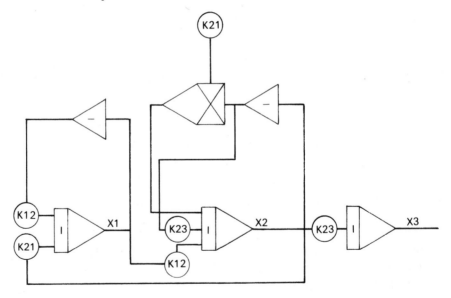

Figure 3-5. 1130/CSMP model of the liver.

3-8
Continuous System Simulation Languages

The approach used in digital-analog simulators is essentially the same as for an analog computer. The user must construct a functional block diagram in which each element is a specific operator. A number of *continuous system simulation languages* have been developed which offer the user greater freedom in describing a system. They make use of a FORTRAN-like statement language, allowing a problem to be programmed directly from the equations of a mathematical model, rather than requiring the equations to be broken into functional elements. As well as providing for addition and integration, they include a variety of algebraic and logical expressions to describe the relations between variables. They, therefore, extend the range of continuous system simulation by removing the orientation towards linear differential equations which characterizes analog methods. Examples of these languages are described in Refs. (4), (5), and (6). One particular language that will be briefly described here is the 360 Continuous System Modeling Program (360/CSMP), (7) and (8).

A 360/CSMP program is constructed from three types of statements:

Structural statements which define the model. They consist of FORTRAN statements and functions, and 34 functional blocks designed for 360/CSMP.

Data statements which assign numerical values to parameters, constants, and initial conditions.

Control statements which specify options for the execution of the program and the choice of output.

Most FORTRAN statements can be used as structural statements. If, for example, the model includes the equation

$$X = \frac{6Y}{W} + (Z - 2)^2$$

the following statement would be used:

$$X = 6.0 \bigstar Y/W + (Z - 2.0) \bigstar \bigstar 2.0$$

Note that the constants are specified in decimal notation. As indicated, it is permissible to include blanks to make the statement easier to read. When punched, the statement can begin in any column. Variable names may have up to six characters.

Also available to the user are 18 of the mathematical function subprograms of FORTRAN. Among them are the exponential function, trigonometric functions, and functions for taking maximum or minimum values. There are also some special CSMP logical functions. Figure 3-6 is a list of 11 of the functional blocks and FORTRAN functions that are available.

GENERAL FORM	FUNCTION
Y = INTGRL (IC, X) Y (0) = IC INTEGRATOR	$Y = \int_0^t X dt + IC$
Y = LIMIT (P_1, P_2, X) LIMITER	$Y = P_1 \qquad X < P_1$ $Y = P_2 \qquad X > P_2$ $Y = X \qquad P_1 \leqslant X \leqslant P_2$
Y = STEP (P) STEP FUNCTION	$Y = 0 \qquad t < P$ $Y = 1 \qquad t \geqslant P$
Y = EXP (X) EXPONENTIAL	$Y = e^X$
Y = ALOG (X) NATURAL LOGARITHM	$Y = \ln (X)$
Y = SIN (X) TRIGONOMETRIC SINE	$Y = \sin (X)$
Y = COS (X) TRIGONOMETRIC COSINE	$Y = \cos (X)$
Y = SQRT (X) SQUARE ROOT	$Y = X^{1/2}$
Y = ABS (X) ABSOLUTE VALUE (REAL ARGUMENT AND OUTPUT)	$Y = |X|$
Y = AMAX1 $(X_1, X_2 \ldots X_n)$ LARGEST VALUE (REAL ARGUMENTS AND OUTPUT)	$Y = \max (X_1, X_2, \ldots, X_n)$
Y = AMIN1 $(X_1, X_2 \ldots X_n)$ SMALLEST VALUE (REAL ARGUMENTS AND OUTPUT)	$Y = \min (X_1, X_2, \ldots, X_n)$

Figure 3-6. 360/CSMP functional blocks.

The following functional block is used for integration:

$$Y = INTGRL(IC, X)$$

where Y and X are the symbolic names of two variables and IC is a constant. The variable Y is the integral with respect to time of X and it takes the initial value IC.

Of the data statements, one called CONST is used to assign numerical values to parameters. Another, called PARAM can also be used to assign values to individual parameters, but its chief purpose is to specify a series of values for one parameter (and only one). The model will be run with the specified parameter taking each of the values on successive runs of the same model. Examples of how these statements are written are

$$CONST A = 0.5, XDOT = 1.25, YDOT = 6.22$$

$$PARAM D = (0.25, 0.50, 0.75, 1.0)$$

As indicated, several values can be specified with each statement by separating the values with commas.

Among the control statements is one called TIMER, which must be present to specify certain time intervals. An integration interval size must be specified. For good accuracy, it should be small in relation to the rate at which variables change value. The total simulation time must also be given. Output can be in the form of printed tables and/or print-plotted graphs. Interval sizes for printing and plotting results need to be specified. The following is an example:

$$TIMER DELT = 0.005, FINTIM = 1.5, PRDEL = 0.1, OUTDEL = 0.1$$

The items specified are

DELT	Integration interval
FINTIM	Finish time
PRDEL	Interval at which to print results
OUTDEL	Interval at which to print-plot

If printed and/or print-plotted output is required, control cards with the words PRINT and PRTPLT are used, followed by the names of the variables to form the output. Several variables can be part of the output and they are listed with their names separated by commas. Two other control cards with the words TITLE and LABEL can be used to put headings on the printed and print-plotted outputs, respectively. Whatever comment is written after the words becomes the heading.

The set of structural data and control statements for a problem can be assembled in any order but they must end with a control card that has the single word END. There are several options for rerunning a problem with changes in some statements but these will not be described. Because of these options, a simple problem must be concluded by following the END card with a STOP card and an ENDJOB card, in that order. Several independent jobs can be run with one loading of the program. When this option is used, the ENDJOB cards (except the last one) are replaced by an ENDJOB STACK card which must be followed by a blank card.

When both data and control statements are punched, the card type name must begin in column 1 and be followed by at least one blank. In the ENDJOB STACK card, the word STACK must begin in column 9.

3-10

360/CSMP Examples

As a first example, Fig. 3-7 shows a program for the automobile suspension problem. It has been coded for the same set of parameter values used in Sec. 3-6. The values of M, K, and F have been specified by a CONST control statement.

```
TITLE   AUTOMOBILE SUSPENSION SYSTEM
*
PARAM   D = (5.656, 14.14, 28.28, 56.56, 70.70)
*
        X2DOT = (1.0/M)*(K*F - K*X - D*XDOT)
         XDOT = INTGRL(0.0,X2DOT)
            X = INTGRL(0.0,XDOT)
*
CONST   M = 2.0, F = 1.0, K = 400.0
TIMER   DELT = 0.005, FINTIM = 1.5, PRDEL = 0.05, OUTDEL = 0.05
PRINT   X, XDOT, X2DOT
PRTPLT  X
LABEL   DISPLACEMENT VERSUS TIME
END
STOP
```

Figure 3-7. 360/CSMP coding for the automobile suspension problem.

The PARAM control statement has been used to program five runs with different values of D. The program integrates with an interval of 0.005 and runs for a time of 1.5. Printed values of X, XDOT, and X2DOT representing, respectively, x, $\dot{x}$, and $\ddot{x}$, and print-plotted output for X have been requested at intervals of 0.05. Figure 3-8 shows the print-plotted graph obtained for the case of $D = 14.14\,(\zeta = 0.25)$.

As a second example, consider the following biological model (9). There are many examples in nature of parasites which must reproduce by infesting some host animal and, in so doing, kill the host. As a result, the population sizes of both the host and parasite fluctuate. As the parasite population grows, the host population declines. Ultimately, the decline in the number of hosts

```
                        MINIMUM            X      VERSUS  TIME        MAXIMUM
                          0.0                                        1.4444E 00
   TIME         X          I                                            I
 0.0          0.0          +
 5.0000E-02   2.1405E-01   -------+
 1.0000E-01   6.8185E-01   --------------------+
 1.5000E-01   1.1389E 00   -----------------------------------+
 2.0000E-01   1.4037E 00   -----------------------------------------------+
 2.5000E-01   1.4266E 00   ------------------------------------------------+
 3.0000E-01   1.2703E 00   -------------------------------------------+
 3.5000E-01   1.0514E 00   -----------------------------------+
 4.0000E-01   8.7691E-01   -----------------------------+
 4.5000E-01   8.0409E-01   --------------------------+
 5.0000E-01   8.3209E-01   ---------------------------+
 5.5000E-01   9.1959E-01   -------------------------------+
 6.0000E-01   1.0135E 00   ----------------------------------+
 6.5000E-01   1.0740E 00   ------------------------------------+
 7.0000E-01   1.0866E 00   ------------------------------------+
 7.5000E-01   1.0604F 00   -----------------------------------+
 8.0000E-01   1.0177E 00   ----------------------------------+
 8.5000E-01   9.8050E-01   --------------------------------+
 9.0000F-01   9.6227E-01   -------------------------------+
 9.5000E-01   9.6471E-01   -------------------------------+
 1.0000F 00   9.8068E-01   --------------------------------+
 1.0500E 00   9.9970E-01   ---------------------------------+
 1.1000E 00   1.0132E 00   ----------------------------------+
 1.1500E 00   1.0173E 00   ----------------------------------+
 1.2000E 00   1.0132E 00   ----------------------------------+
 1.2500E 00   1.0050E 00   ---------------------------------+
 1.3000E 00   9.9722E-01   --------------------------------+
 1.3500E 00   9.9287E-01   --------------------------------+
 1.4000E 00   9.9270E-01   --------------------------------+
 1.4500E 00   9.9553E-01   --------------------------------+
 1.5000E 00   9.9932E-01   --------------------------------+
```

Figure 3-8. Print-plot output for the automobile suspension problem.

causes a decline in the birth of parasites and, consequently, the population of the hosts begins to climb. The process can continue to cause oscillations indefinitely.

To construct a model of the system, let X be the number of hosts and Y be the number of parasites. Suppose the excess of the birth rate of the hosts over the death rate from natural causes is A (assumed positive). In the absence of parasites, the population of the hosts should grow according to the equation

$$\dot{X} = AX$$

The death rate from infection by the parasites depends upon the number of encounters between the parasites and hosts, which is assumed to be proportional to the product of the numbers of parasites and hosts. The equation controlling the growth of the hosts is, therefore,

$$\dot{X} = AX - KXY$$

We will make the simplifying assumption that each death of a host due to a parasite results in the birth of one parasite. This is the only means by which the parasite population can grow, but the parasites are subject to a natural death rate of D. Hence the equation controlling the parasite population is

$$\dot{Y} = KXY - DY$$

Figure 3-9 shows a program for solving the two simultaneous equations for the particular values

$$A = 0.005$$

$$K = 6 \times 10^{-6}$$

$$D = 0.05$$

```
TITLE   PARASITE-HOST MODEL
*
        XDOT = A*X - K*X*Y
        YDOT = K*X*Y - D*Y
        X = INTGRL(XO,XDOT)
        Y = INTGRL(YO,YDOT)
*
CONST   A = 0.005, D = 0.05, XO = 10000.0, YO = 1000.0, K = 0.000006
*
TIMER   DELT = 0.1, FINTIM = 500.0, PRDEL = 10.0
PRINT   X,Y
END
STOP
```

Figure 3-9. Program for the host-parasite problem.

The initial values of X and Y were 10,000 and 1,000, respectively. Figure 3-10 plots the output of the program.

Exercises

Derive a CSMP program for the following problems.

3-1 Modify the automobile suspension program of Sec. 3-6 so that the displacement of the mass cannot exceed 1.25.

3-2 Reprogram the automobile suspension problem with the assumption that the damping force of the shock absorber is equal to $5.656(\dot{x} - 0.05\ddot{x})$.

3-3 Solve the following equations:

$$3\dddot{x} + 15\ddot{x} + 50\dot{x} + 200x = 10$$

$$\dddot{x} = \ddot{x} = \dot{x} = x = 0 \qquad \text{at } t = 0$$

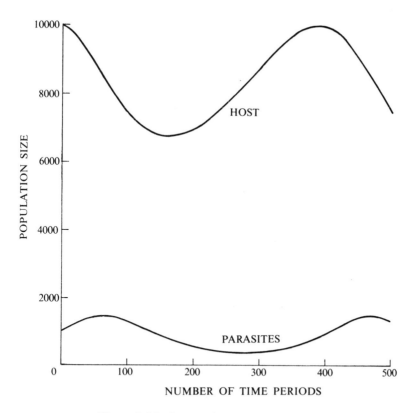

Figure 3-10. Results of host-parasite problem.

3-4 A torsional pendulum consists of a bar suspended from a spring that winds and unwinds as the bar oscillates up and down. If x is the vertical displacement and θ is the angle the spring turns, the following equations describe the motion:

$$a\ddot{x} + bx - c\theta = F(t)$$
$$e\ddot{\theta} + f\theta - gx = G(t)$$

Assume the system starts from rest with x and θ equal to zero, $F(t)$ remains zero for all time and $G(t)$ is 1 for $t \geqslant 0$. Solve the equations for the following values:

$$a = e = 1$$
$$b = f = 40$$
$$c = g = 4$$

3-5 In a chemical reaction one molecule of a substance X is produced for one molecule each of substances A and B. The initial concentrations of A and B are a and b, respectively. Let x be the concentration of X and assume that it is initially zero. The rate at which x increases is 0.1 times the product of the current concentrations of A and B. Assume a and b are initially 0.8 and 0.4, respectively, and simulate the production of X.

3-6 The following model has been proposed to describe the population growth of a species of animal living in a confined space. The excess of births over natural deaths causes a growth rate of a times the current number of individuals, N. Competition for food causes death from starvation at a rate bN^2. Simulate the population growth assuming $a = 0.05$, $b = 1 \times 10^{-5}$, and $N = 500$ at zero time.

3-7 Assume in the previous problem that a second species exists in the same space and is subject to the values $a_1 = 0.04$, $b_1 = 1.5 \times 10^{-5}$, and $N_1 = 1,000$ at time zero. However inter-species fighting causes deaths to both species at a rate that is 1×10^{-6} times the product of their numbers. Simulate their population growths.

Bibliography

1 Warfield, John N., *Introduction to Electronic Analog Computers*, Englewood Cliffs, N. J.: Prentice-Hall, Inc., 1959.

2 Linebarger, R. N., and R. D. Brennan, "A Survey of Digital Simulation: Digital-Analog Simulation Programs," *Simulation*, III, No. 6 (Dec., 1964), 22–26.

3 1130 Continuous System Modeling Program, Form No. H 20-0282, IBM Corp., Data Processing Division, White Plains, N.Y.

4 Syn, W. M., and R. N. Linebarger, "DSL/90—A Digital Simulation Program for Continuous System Modeling," *AFIPS Conference Proceedings*, *SJCC.*, Vol. XXVIII, Washington, D. C.: Spartan Books, (April, 1966) 165–187.

5 Sansom, F. J., and H. E. Peterson, "MIMIC—A Digital Simulation Program" SESCA Internal Memo. 65–12, Wright-Patterson Air Force Base, Ohio, May, 1965.

6 Schlesinger, S. I., and L. Sashkin, "EASL—A Digital Computer Language for 'Hands-On' Simulation," *Simulation*, VI, No. 2 (Feb., 1966).

7 Brennan, R. D., and M. Y. Silberberg, "Two Continuous System Modeling Programs," *IBM Systems Journal*, VI, No. 4 (1967), 242–266.

8 System/360 Continuous System Modeling Program, Form No. H 20-0367, IBM Corp., Data Processing Division, White Plains, N.Y.

9 Volterra, V., *Leçons sur la Théorie Mathématique de la Lutte pour la Vie*, Paris: Gauthier-Villars, 1931.

Lessons about the mathematical theory of the struggle for life.

4

INDUSTRIAL DYNAMICS

Industrial Dynamics aims to study the performance of business corporations or entire industries by using simulation techniques to show how they respond to various conditions. The term has its origin in work done at the Massachusetts Institute of Technology by J. W. Forrester and his co-workers (1). For an historical discussion of Industrial Dynamics and references to several applications, see (2), (3), and (4).

The operation of a business is carried out by performing a number of functions such as production, distribution, marketing, and financing. These functions are identified as the activities of a system and such elements as men, materials, money, orders, equipment, and information are treated as the system entities. To describe the behavior fully, it is necessary to consider a business as a whole, so that all the interactions are seen. Typically, an Industrial Dynamics study will cover the production, distribution, and retailing of a commodity in order to incorporate all the components of the business system.

The prime objective of an Industrial Dynamics study is to understand how the organization of a business will affect its performance. Simulation techniques are used to test alternative management policies. The simulation aims to demonstrate the characteristic behavior of the system rather than to predict specific events. The individual events in a system, such as the placing of an order or the shipping of a product, are discrete events. An Industrial Dynamics study, however, aggregates the events and treats the system as a continuous system.

An Industrial Dynamics view of a system concentrates on the rates at which various quantities change and expresses the rates as continuous variables. The flow of orders from customers will be described as a certain rate of orders and the delivery of orders will similarly be represented by a rate. The difference between the two rates will be integrated to give the level of unfilled orders at any time.

4-2

Industrial Dynamics Diagrams

The basic structure of an Industrial Dynamics model is illustrated in Fig. 4-1. It consists of a number of reservoirs, or levels, interconnected by flow paths. The rates of flow are controlled by decision functions that depend upon conditions in the system. The *levels* represent the accumulation of various entities in the system such as inventories of goods, unfilled orders, number of employees, etc. The current value of a level at any time represents the accumulated difference between the input and the output flow for that level. *Rates* are defined to represent the instantaneous flow to or from a level. *Decision functions* or, as they are also called, *rate equations* determine how the flow rates depend upon the levels.

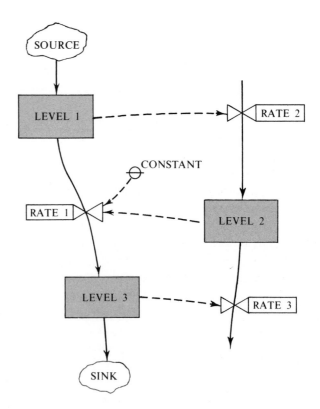

Figure 4-1. Structure of an Industrial Dynamics model.

A set of symbols has been established for indicating the various factors involved in an Industrial Dynamics model and some of these are illustrated in Fig. 4–1. Levels are represented by boxes; and rates are indicated by a symbol which, as a matter of interest, is adapted from the symbol used in diagrams of hydraulic systems to represent a valve. It is sometimes necessary to indicate a source or sink of entities, as is shown in Fig. 4-1. For completeness, the need for a constant in the model is indicated with a special symbol. A full Industrial Dynamics model will show the flow of many different types of entities by different types of lines. For this simplified explanation, however, it is not necessary to draw these distinctions.

Levels would appear to be dimensionless quantities since they represent a count. However, the term level is also applied to some quantities that do have dimensions. For example, the rate of reordering goods may depend upon the average number of orders over some period of time, which is a quantity that has the dimensions of number over time. Nevertheless, the average number of orders per week would be regarded as a level in an Industrial Dynamics model. The simplest test for determining which quantities are to be regarded as levels is to imagine the system brought to rest. Any quantity that is a rate is then automatically zero, while any quantity that maintains a magnitude should be regarded as a level.

4-3
A Simple Industrial Dynamics Model

Mathematically, a rate is represented by the derivative of a variable. Since an Industrial Dynamics model relates levels to rates it follows that an Industrial Dynamics model is actually a set of differential equations. Usually, the model is constructed from simple linear differential equations with constant coefficients. To illustrate this point, consider the following system.

A builder observes that the rate at which he can sell houses depends upon the number of families who do not yet have a house. As the number of people without houses diminishes, the rate at which he sells houses drops.

Let y_0 be potential number of house-owning families,

y be number of families with houses.

The situation is represented in Fig. 4-2. The horizontal line at y_0 is the total potential market for houses. The curve for y indicates how the number of houses sold increases with time. The slope of the curve (i.e., the rate at which y increases) decreases as y_0-y gets less. This reflects the slowdown of sales as the market becomes saturated. Mathematically, the trend can be expressed by the equation:

$$\dot{y} = K_1(y_0 - y)$$
$$y = 0 \quad \text{at } t = 0$$

(4-1)

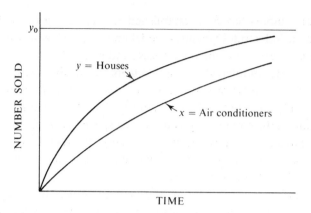

Figure 4-2. Sale of houses and air conditioners.

Consider now a manufacturer of central air conditioners designed for houses. His rate of sales depends upon the number of houses built. (For simplicity, it is assumed all houses will install an air conditioner.) In the same way as for house sales, the rate of sales diminishes as the unfilled market diminishes.

Let x = number of installed air conditioners. Then the unfilled market is the difference between the number of houses and the number of installed air conditioners. The sales trend may be expressed mathematically by the equation

$$\dot{x} = K_2(y - x) \tag{4-2}$$

The change of x with time is also illustrated in Fig. 4-2.

Solving Eq. (4-2) for y and substituting in Eq. (4-1) and then also differentiating the expression for y derived from (4-2) and substituting for $\dot{y}$, it will be found that the combined result of the two trends on the number of installed air conditioners is represented by the equation

$$\ddot{x} + (K_1 + K_2)\dot{x} + K_1 K_2 x = K_1 K_2 y_0$$

Now assume that the potential number of house-owning families y_0 is not static but is a function of time $F(t)$. Then, the equation determining the number of air conditioners sold is

$$\ddot{x} + (K_1 + K_2)\dot{x} + K_1 K_2 x = K_1 K_2 F(t) \tag{4-3}$$

It can immediately be seen that Eq. (4-3) is a second-order linear differential equation with constant coefficients of exactly the same form as Eq. (3-1) which describes a mechanical system.

When discussing the motion of an automobile wheel, in Sec. 1-7, it was pointed out that the solution of Eq. (3-1) can be oscillatory. In the particular form in which Eq. (4-3) appears it will be found that, as long as K_1 and K_2 are positive, it is impossible for the oscillatory conditions to occur.

A typical use that might be made of the model just derived is to estimate the effect of a change in mortgage interest rates on business. A sudden change in rates will alter the potential house-owner market y_0 by a step function. The response to the change can be examined with the model, just as the motion of an automobile wheel in response to a sudden applied force can be examined by Eq. (3-1).

4-4

Representation of Delays

A very significant factor in determining the response of an industrial system is the delay that occurs in transmitting goods and information and in making decisions. Some delays are fixed intervals of time. However, the aggregation of many individual discrete events into a continuous representation practiced in Industrial Dynamics requires a correspondingly "smooth" representation of delays.

The method used in Industrial Dynamics to represent a delay is to control the rate at which a level can change by the level that causes the change. Suppose, for example, a factory is shipping goods and at time $t = 0$ has X_0 outstanding orders. Let x be the number of outstanding orders at time t. The level that controls the rate in this case is x itself, so that,

$$\dot{x} = -\frac{1}{T}x$$

(4-4)

$$x = X_0 \qquad \text{at } t = 0$$

The negative sign shows that x is decreasing with time. The solution of the equation for x is

$$x = X_0 e^{-t/T}$$

Curve C of Fig. 3-1 plots x as a function of t/T, showing how the level of outstanding orders diminishes with time. The constant T controls the delay. Smaller values of T increase the rate $\dot{x}$ and result in less delay. It can, in fact, be shown that T is the average delay.

Consider now the number of goods delivered, denoted by y. The variable y increases from 0 to X_0 in such a way that $x + y = X_0$. The level that determines the rate at which y changes is the number of undelivered goods, so that

$$\dot{y} = \frac{1}{T}(X_0 - y)$$

$$y = 0 \qquad \text{at } t = 0$$

(4-5)

The solution is

$$y = X_0(1 - e^{-t/T})$$

which is plotted as curve B of Fig. 3-1.

The general formula used for expressing a delay is

$$\text{Rate} = \frac{\text{Level}}{\text{Delay}}$$

(4-6)

The two equations describing the air conditioner sales model, Eqs. (4-1) and (4.2), have exactly the same form as Eq. (4-5). It can now be seen that the relationships can be interpreted by saying that there is a delay $T_1 = 1/K_1$ between the potential housing market and the sale of houses and another delay $T_2 = 1/K_2$ between house sales and air conditioner sales.

The corresponding Industrial Dynamics block diagram is shown in Fig. 4-3. There are two levels to represent the number of houses sold and the number of air conditioners sold, and two rates to control how each level increases with time. The rates depend upon the levels and the constants of the system, and this is shown by the lines leading to the rate symbols.

Both these rates represent the fact that there is a delay in building up a level. The same general method is used to indicate a delay in reducing a level.

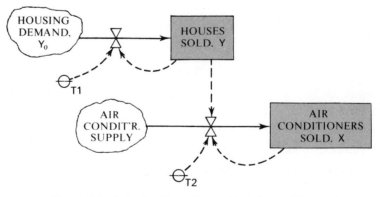

Figure 4-3. Industrial Dynamics model of air conditioner sales.

Suppose, for example, we wish to represent the fact that air conditioners break down after a while and so some house owners become potential customers again. Let T_3 be the average life of an air conditioner. Then a third rate must be introduced to decrease the level of x. The corresponding Industrial Dynamics model is shown in Fig. 4-4.

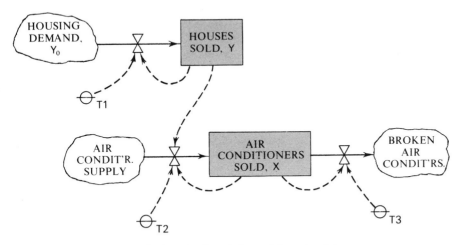

Figure 4-4. Introduction of air conditioner breakdowns.

Because the delays in this case are represented by a first-order differential equation, they are referred to as first-order delays. Consider now the over-all delay between potential housing market y_0 and air conditioner sales x. As was demonstrated, the relationship between these quantities is a second-order differential equation. This is said to be a second-order delay of magnitude $T_1 + T_2$. Similarly, the relationship between potential housing market and broken-down air conditioners is a third-order delay corresponding to all three blocks of Fig. 4-4. If desired, higher order delays could be defined.

The transient nature of the response changes with the order of the delay, so that, if a first-order delay is not considered to be satisfactory, a better representation can sometimes be obtained by using a higher order delay. For example, a delay of size T could be two first-order delays of size $T/2$. As a rule, however, the level of abstraction being used in an Industrial Dynamics model does not justify this degree of accuracy. For the remaining examples used here, a delay will be interpreted as a first-order differential equation.

4-5

Feedback Systems

A particularly significant factor in the performance of many systems is that a coupling occurs between the input and output of the system. The term *feedback* is used to describe the phenomenon of having the output of a system affect the input to the system. As might be expected, the feedback can greatly affect the behavior of a system. Depending upon how the feedback is controlled, the effect can be desirable or disastrous.

A home heating system controlled by a thermostat is a simple example of a feedback system. The system has a furnace whose purpose is to heat a room, and the output of the system can be measured as room temperature. Depending upon whether the temperature is below or above the thermostat setting, the furnace will be turned on or off, so that information is being fed back from the output to the input. In this case, there are only two states, either the furnace is on or off.

An example of a feedback system in which there is continuous control is the aircraft system discussed previously in Sec. 1-1 and illustrated in Fig. 1-1. Here, the input is a desired aircraft heading and the output is the actual heading. The gyroscope of the autopilot is able to detect the difference between the two headings. A feedback is established by using the difference to operate the control surfaces, since change of heading will then affect the signal being used to control the heading. The difference between the desired heading θ_1 and actual heading θ_0 is called the error signal ε since it is a measure of the extent to which the system deviates from the desired condition.

Suppose the control surface angle is made directly proportional to the error signal. The force changing the heading is then proportional to the error and, consequently, it diminishes as the aircraft approaches the correct heading. Consider what happens when the aircraft is suddenly asked to change to a new heading θ_1. The subsequent changes are illustrated in Fig. 4-5. The upper curves represent the control surface angle and the lower curves represent the aircraft heading.

Consider first the solid curves of Fig. 4-5. Upon receipt of a signal to change direction, the error signal, and therefore the control surface angle, suddenly takes a non-zero value. The aircraft heading, shown in the lower curves of Fig. 4-5, responds to the control surface by moving toward the new heading but, because of inertia, it takes time to respond. As the aircraft turns, the control surface angle decreases so that less force is applied as the aircraft approaches the required heading. Ultimately, when the aircraft reaches the required heading, the control surface angle will be zero but the inertia of the aircraft can carry it beyond the desired heading. As a result, the control surface turns in the opposite direction in order to bring the aircraft back from its over-shoot. The correction from the overshoot produces an undershoot and the motion follows a series of oscillations of decreasing amplitude as illustrated in Fig. 4-5.

Suppose the system is changed so that for the same size error signal there is a larger control surface angle. The dotted curves of Fig. 4-5 represent this case. For the same conditions, the restoring force is larger and the aircraft responds

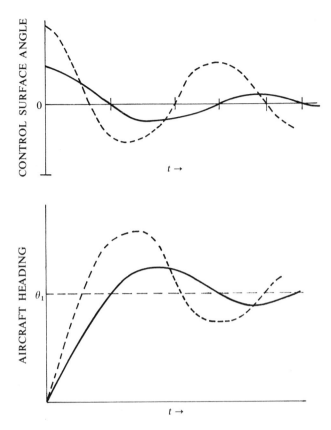

Figure 4-5. Aircraft response to autopilot system.

more rapidly but, correspondingly, the initial overshoot will be larger. The aircraft will oscillate more widely and rapidly, as is illustrated in Fig. 4-5. The feedback loop in the second case is said to have greater *amplification*, since the same error produces a larger correction force. Under certain conditions, it is possible for the amplification to be so great that the initial overshoot exceeds the initial error; in which case, the undershoot from the correction becomes even larger and the system becomes unstable because of ever-increasing oscillations.

Changes in the characteristics of the aircraft will cause changes in the response. The sluggishness of the aircraft can be interpreted by saying that the response involves a delay. The response is also affected by any other delay that occurs in the feedback loop, such as a delay in measuring the error signal. It can be seen, therefore, that when feedback occurs, the over-all response of the system is affected by the amplification and delays that occur in the system.

In the autopilot system just discussed, an increase in the system output *decreases* the input. Such systems are said to have *negative* feedback. The feedback is said to be *positive* if an increase of the output *increases* the error signal. Generally, positive feedback is avoided because it tends to make systems unstable. For example, the aircraft response can be speeded up by adding to the error signal a term proportional to the rate at which the aircraft is turning. Initially, this signal is a positive feedback because the rate of turning increases as the error decreases. If this component of the control signal is made too large, it will force the aircraft into ever-increasing oscillations. A similar situation can occur in a business. Suppose, for example, a company invests money in advertising and the resulting sales increase more than covers the cost. If the advertising budget is tied directly to sales volume, the sales increase results in yet more money for advertising, which can cause an explosive rate of expansion.

It is not to be concluded from this simple discussion of feedback systems that negative feedback systems are necessarily stable or that positive feedback systems are necessarily unstable. It has already been seen that excessive amplification in a negative feedback system can cause instability; it will be found that excessive delays can also cause unstable conditions. In the advertising example which had positive feedback, the effects that caused an explosive expansion in the sales can cause an equally dramatic collapse if the trend in sales decreases.

The examples discussed here have necessarily been simple. In practice, where feedback occurs there may be many interconnected feedback loops and the behavior of the loops is complicated by non-linear effects. Many simulation studies of continuous systems are concerned primarily with analyzing the response of the system and striking a balance between a fast response and the need to maintain satisfactory stability.

4-6
Feedback in Industrial Systems

Feedback occurs in industrial systems in much the same way it occurs in physical systems. The feedback is not usually caused by a physical signal, but occurs through the feedback of information. Some decision is made on the basis of information about prevailing conditions, and the action taken as a result of the decision changes the conditions upon which the information is based. The link between the decision and the information may go unnoticed. There could be a long chain of circumstances connecting the two. There has, for example, been much speculation that feedback causes business cycles in the economic growth of a country and many efforts have been made to identify the underlying forces. On the other hand, feedback may deliberately be introduced in an industrial system to help control the system. A specific example will be discussed in the next section.

The delays in an industrial system may be caused by production or transportation times, or they may simply be delays in acquiring accurate information.

It is difficult, for example, to keep records of inventory levels up-to-date so that any decision based on the inventory level is based on old information and, in effect, is subject to a delay. The amplification that occurs in an industrial feedback loop is controlled by the decisions being made in the system. A buyer who orders new stock on the basis of an inventory level controls the amplification by deciding how many items he orders in relation to the drop in inventory level.

Most Industrial Dynamics studies are concerned with studying the effects of feedback on system performance. This explains why the technique used simplifies the model by regarding the changes in the system as occurring continuously. The principal objective is to study stability and decide what management decisions can best maintain stability and yet give a reasonably fast response.

4-7

Inventory Control Systems

A feedback system that is analogous to the autopilot system is seen in the control of inventories. A retailer will usually maintain an inventory of goods from which to make sales. He aims to keep the inventory at a level that strikes a reasonable balance between the cost of holding goods in inventory and the penalty of losing sales if the inventory should become empty.

Figure 1-1, which represents the autopilot system, could also represent an inventory control system. The input θ_1 that previously represented desired aircraft heading represents the desired inventory level. The output θ_0 that represented actual aircraft heading now represents actual inventory level. The difference or, as it was previously called, error signal controls the orders placed to replenish the inventory. The delay, represented by the inertia of the aircraft, is the delay in filling the orders. In a manner analogous to the aircraft heading, the inventory level can oscillate and it is even possible that, if the retailer overreacts by ordering large quantities for small differences, the system corresponds to the autopilot system with too much amplification and will be unstable.

In practice, an inventory system does not usually have a negative error signal; that is to say, when the inventory exceeds the desired level, the orders do not go negative. If they did, it would correspond to saying that goods are sent back. It is more likely that there is a steady drain on the inventory which corresponds to normal sales. The retailer establishes a steady average order *rate* which balances the average rate of sales and maintains a constant inventory level. The changes represent changes with respect to the average order rate, so that a negative error signal corresponds to reducing the order rate below the normal amount.

4-8

Industrial Dynamics Model of an Inventory Control System

Consider now an Industrial Dynamics model of the inventory control system just described. A simple model of the inventory control system would need two levels and three rates defined as follows:

X Current inventory level
Y Outstanding level of orders placed with the supplier
U Rate of ordering from supplier
V Rate of delivery from supplier
S Rate of sales

In addition, two constants need to be defined:

I.   Planned inventory level
T_1 Average delivery time

A model of the system is shown in Fig. 4-6. It shows the level of outstanding orders being controlled by the rate at which orders are placed and the rate of delivery. Similarly, the level of the inventory is controlled by the rate of delivery and the rate of sales.

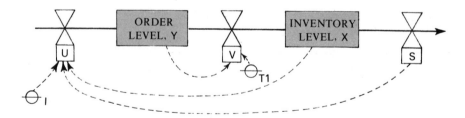

Figure 4-6. Industrial Dynamics model of an inventory control system.

To derive a mathematical model of the system, equations can immediately be written for the two levels. These are

$$\dot{Y} = U - V \tag{4-7}$$

$$\dot{X} = V - S \tag{4-8}$$

Using the general formula for delays, the equation for the rate of delivery can be written

$$V = \frac{Y}{T_1} \tag{4-9}$$

The rate of ordering is a management decision. Clearly, it should balance the steady sales rate, but simply to equate order rate to sales rate is not satisfactory. When the sales rate increases, the inventory will become depleted because the delivery rate takes time to catch up with the new sales rate. Similarly, a drop in sales rate will result in the inventory level rising. The decision, therefore, is to add to the sales rate a term proportional to the difference between the desired level and the actual level. When the inventory is low, this increases the order rate and, when the level is high, it decreases the rate. The order rate is, therefore,

$$U = S + K(I - X) \tag{4-10}$$

where K is a constant that will be called the *ordering constant.*

This ordering policy introduces feedback. It is apparent that the value of K will affect the system response, since K corresponds to the amplification in the feedback loop. Its value, along with the value of T_1, will determine if the system oscillates or not.

The set of Eqs. (4-7) to (4-10) describes the system completely, and they can be solved analytically; however, a 360/CSMP program for their solution is shown in Fig. 4-7. Assume that time is measured in units of days and let the system be initialized so that, at time $t = 0$,

$$I = X = 20$$

$$Y = 12$$

$$S = 4 \text{ items/day}$$

Let the average delivery delay T_1 be 3 days. Suppose the demand suddenly increases at time $t = 4$ to 6 items/day. The changes in inventory level for

```
TITLE   SIMPLE INVENTORY CONTROL SYSTEM
*
PARAM   K = (1.0, 0.5, 0.25, 0.125, 0.0625)
*
        S = 4.0 + 2.0*STEP(4.0)
        V = Y/T1
        U = S + K*(I - X)
        YDOT = U - V
        Y = INTGRL(Y0,YDOT)
        XDOT = V - S
        Y = INTGRL(X0,XDOT)
*
CONST   I = 20.0, Y0 = 12.0, XC = 20.0, T1 = 3.0
*
TIMER   DELT = 0.1, FINTIM = 50.C, PRDEL = 1.0
PRINT   U,V,X,Y
END
STOP
```

Figure 4-7. 360/CSMP program for an inventory control system.

various values of K are shown in Fig. 4-8. It can be seen that oscillations can occur. The ordering policy that has been adopted does, however, achieve its objective since the inventory settles to its planned level.

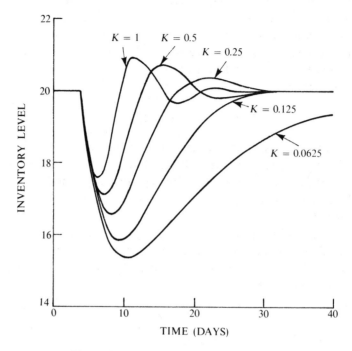

Figure 4-8. Inventory control system response.

The oscillations are more pronounced when $K = 1$. They diminish as K decreases, and it can be shown that they disappear when $K = \frac{1}{12}$. It should be remembered that the term $K(I - X)$ has been added to the ordering *rate*. It is not a one-time order intended to replace a deficit detected in the inventory level. Some insight can then be gained as to why large values of K cause oscillations. With $K = 1$, for example, enough goods are shipped in a day to fill the deficit but, since there is a three-day delay in delivery, the term added to S will then reorder enough to fill the deficit several times over.

It would, of course, be possible to have an inventory control system based on the policy that, periodically, one-time orders are placed to wipe out accumulated inventory deficits. However, it is important to have some methodical way of correcting deviations automatically, particularly where demand fluctuates a great deal. The rule that has just been tested is a reasonably good rule provided that K is chosen to be a value for which there is little or no oscillation. A suitable value in this simple system is about $\frac{1}{2}$. The value will, of course, change with the delay time T_1. Many other inventory control rules could be used, and it requires a considerable amount of study to select a rule best suited for any particular system and type of demand (5).

The cases investigated did not result in the inventory level going negative at any time. In general, however, a mathematical model may allow negative values to occur. A complete model should, therefore, include limits to prevent values that by their nature cannot fall below zero from going negative.

There are two implicit assumptions in the simple model that has been discussed. If the inventory level falls to zero, the system will be unable to supply goods to customers. Since no record is being kept of the level of customer orders, it is not possible to make up the orders. The model corresponds to the case where sales would be lost in these circumstances. It is also being assumed that the retailer's orders for new stock are always filled so that it is being implied that the supplier has a large enough stock to meet all demands.

To remove these restrictions, it is necessary to distinguish between the flow of orders and goods, thereby producing a model that is more characteristic of an Industrial Dynamics study. We now define four levels in the model to represent both the backlog of orders and inventory for the retailer and his supplier. A flow chart of the system is shown in Fig. 4-9.

As before, X represents the level of inventory held by the retailer and Y is the level of orders placed with the supplier. The levels W and Z are introduced to represent the orders held by the retailer and the inventory held by the supplier. The rates, U, V, and S, as before, represent, respectively, the rate at which orders are sent to the supplier, the rate at which he supplies goods, and the rate at which customers order goods from the retailer. Two new rates are D, the rate at which goods are delivered to customers, and R, the rate at which the supplier receives goods from his source of supply.

The flow chart shows orders flowing into the retail backlog level W at a rate S. The rate of delivery to the customers, D, depletes the retail inventory level X and simultaneously removes orders from the backlog W. Similarly, there is a flow of orders at a rate U into the supplier's backlog level Y, and his delivery rate V depletes both the inventory level Z and the backlog level Y. The supplier's inventory is increased at a rate R. Whereas, before, the delivery rate to customers depended upon the retailer's inventory level X, the delivery rate D will more realistically be based upon the retailer's level of orders. A delay of T_2 will be assumed between the receipt of orders by the supplier and his delivery of goods.

In writing a program for the improved model, limiters will be introduced to stop the inventories from going negative. In addition, a more compact way of writing the level equations will be used. When using the INTGRL operator

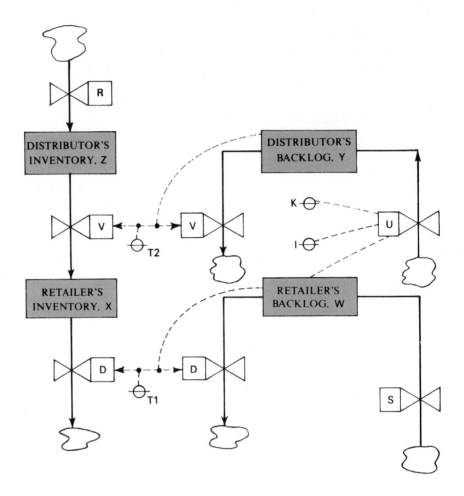

Figure 4-9. Improved inventory control system.

of 360/CSMP, it is not necessary that the integrand be a simple variable; it can be an expression. In fact, it can be another INTGRL operator so that one statement can perform a multiple integration. (However, this does not save computing time.) The expression for the rates at which the levels change will be inserted directly in the INTGRL operators, reducing the size of the model.

Figure 4-10 is a 360/CSMP program for the improved inventory control system model. The symbols X1 and Z1 have been introduced for the outputs of the inventory level equations. They are both modified by limiters to produce the actual inventory levels X and Z.

Figure 4-11 shows outputs for the particular case where $K = 0.5$. As before, the value of T_1 is 3 days but T_2 has been set at 4 days. The rate at which the supplier receives goods, R, has been set to a constant value of 4 items/day, and

```
TITLE    EXPANDED INVENTORY CONTROL SYSTEM
*
PARAM    K = (1.0, 0.5, 0.25, 0.125, 0.0625)
*
         S  = 4.0 + 2.0*STEP(4.0)
         V  = Y/T2
         D  = W/T1
         U  = S + K*(I - X)
         X1 = INTGRL(X0,(V - D))
         X  = LIMIT(0.0,1000.0,X1)
         Y  = INTGRL(Y0,(U - V))
         Z1 = INTGRL(Z0,(R - V))
         Z  = LIMIT(0.0,1000.0,Z1)
         W  = INTGRL(W0,(S - D))
*
CONST    I=20.0,X0=20.0,Y0=16.0,W0=12.0,Z0=100.0,R=4.0,T1=3.0,T2=4.0
*
TIMER    DELT = 0.1, FINTIM = 50.0, PRDEL = 1.0
PRINT    X,Y,Z,W
END
STOP
```

Figure 4-10. Program for extended inventory control system.

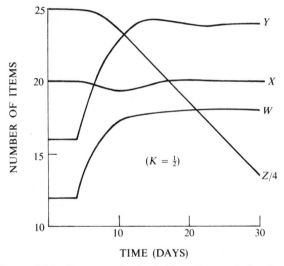

Figure 4-11. Response of extended inventory control system.

his initial inventory level has been set to 100. He is not using the inventory control rule of the retailer, so his stock continually diminishes because of the increase in demand.

Exercises

4-1   Program the air conditioner sales model of Sec. 4-3, allowing for break-downs. Assume the average time to sell a house is 5 months, the average time to install an air conditioner is 10 months and breakdowns occur, on the average, after 25 months. Take the initial housing market to be 1,000 houses.

4-2 In the model of air conditioner sales, without breakdowns, that is given in Sec. 4-3, prove that the sales response cannot be oscillatory if K_1 and K_2 are positive. (Use the relation $(K_1 - K_2)^2 \geq 0$.)

4-3 In the aircraft system discussed in Sec. 4-5, suppose the force applied by the control surface is 160 times the error signal. Suppose, further, that the response of the aircraft to the applied force is given by

$$10\ddot{\theta}_0 + 60\dot{\theta}_0 = \text{Force}$$

With what frequency and damping will the aircraft oscillate?

4-4 In the preceding problem, assume a signal equal to $A\dot{\theta}_0$ is added to the error signal. What is the frequency and damping when $A = 0.1$? What values of A will make the aircraft unstable?

4-5 Babies are born at the rate of 1 baby per annum for every 20 adults. After a delay of 6 years they reach school age. Their education takes 10 years, after which they are adults. The adults die after an average adult life of 50 years. Draw an Industrial Dynamics diagram of the population and program the model assuming the initial numbers of babies, school-children, and adults are, respectively, 300, 3,000, and 100,000.

4-6 Extend the model of Sec. 4-9 further to include a segment showing the supplier ordering his goods from a wholesaler. The supplier uses the same inventory control policy as the retailer, and there is a four-day delay in delivery from the wholesaler.

4-7 Assume the supplier in the model of Sec. 4-9 has a factory that manufactures the goods he supplies. He sends orders for manufacturing goods at the same rate as he receives orders. There are three successive one-day delays involved in the manufacturing, and the factory output goes to the supplier's inventory. Draw a diagram and program the problem.

Bibliography   1 Forrester, Jay W., *Industrial Dynamics*, Cambridge, Mass.: The M.I.T. Press, 1961.

2 Forrester, Jay W., "Industrial Dynamics—After the First Decade," *Management Science*, XIV, No. 7 (March, 1968), 398–415.

3 Forrester, Jay W., "New Directions in Industrial Dynamics," *Industrial Management Review*, VI, No. 1, Alfred P. Sloan School of Management, M.I.T. (Fall, 1964).

4 Ansoff, H. Igor, and Dennis P. Slevin, "An Appreciation of Industrial Dynamics," *Management Science*, XIV, No. 7 (March, 1968), 383–397.

5 Plossl, George W., and Oliver W. Wight, *Production and Inventory Control— Principles and Techniques*, Englewood Cliffs, N. J.: Prentice-Hall, Inc., 1967.

5

THE DYNAMO
PROGRAMMING
LANGUAGE

As has been shown in Chap. 4, the technique employed in Industrial Dynamics is to construct a mathematical model of a system, consisting mainly of first-order differential equations. With a model of this form, a number of computational methods could be used to solve the equations, and examples were given of Industrial Dynamics models being solved with the simulation language 360/CSMP. One particular programming language that will now be described was specifically developed for Industrial Dynamics models. The language called DYNAMO (for Dynamic Models) was developed by A. Pugh and his co-workers at M.I.T. (1).

The flow chart symbols and flow charting method described in Sec. 4-2 and illustrated in Fig. 4-1 were in fact developed for use with DYNAMO. Variables in DYNAMO are represented by symbols of from one to five characters, and it is customary to mark the level and rate symbols of the block diagram with the symbols to which they refer, as was done in Fig. 4-1.

The way time is represented is illustrated in Fig. 5-1. The name TIME is reserved for making reference to the time in the system model; the instant at

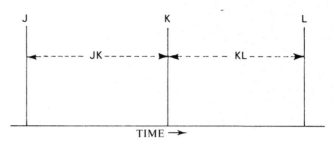

Figure 5-1. Representation of time in DYNAMO.

which calculations are being made is referred to as TIME.K. The previous instant at which calculations were made is TIME.J and the next instant at which calculations will be made is TIME.L. The interval just passed is called the JK interval, and the interval coming up is the KL interval. Calculations are made for uniform intervals of time, so that the JK and KL intervals are always the same size. The length of the constant interval is designated by the symbol DT. The size of DT is chosen by the user; the criteria for making the choice are discussed in the next section.

The convention adopted is to attach one of the symbols J, K, L, JK, or KL as a suffix to a variable. When cards are punched, the name of the variable is followed by the suffix separated by a period. Thus, a level called INV at time J will be indicated by INV.J, and the value of a rate INP during the interval JK will be indicated by INP.JK. Where a symbol is to represent a constant, it is not necessary to attach any suffix.

5-3

Choice of Solution Interval

The value of DT needs to be chosen individually for each model. The action of the program is to compute the values of all variables at successive intervals of DT time units. If DT is too large, the results become inaccurate, because they do not follow closely enough the continuous interaction between the variables; the solution of the equations can, in fact, become unstable. As the value of DT is made smaller, the solution accuracy is increased at the expense of increasing the amount of computer time spent on calculation. No precise rule can be given for determining the point at which a good balance between these two factors is obtained. In practice, it is customary to try a number of different values of DT, to determine whether the results are significantly affected by the choice of value.

Changes in the system do not occur instantaneously because of the delays in the system. The length of DT, therefore, should be judged in relation to the magnitude of the delays. As a general rule of thumb, DT should be no bigger than half the smallest delay time in the system. In applying this rule, it must be remembered that an Nth order delay of magnitude T is in fact a series of N first order delays, each of magnitude T/N. Therefore, the value of DT should be at least as small as $T/2N$.

It is preferable that DT be smaller than this value, because the accuracy will be poor for those variables depending on the smallest delay. Depending upon the significance of these variables, the over-all results may be poor. A useful practice is to use the rule of thumb in the period when the model is being constructed and the general nature of the response is being investigated, and to reduce the interval size when results are being taken for the system study. This subject is discussed more fully in Appendix D of Ref. (3).

5-4

Equation Forms

To simplify programming, DYNAMO defines a number of equation forms, each of which is a prototype. All equations must comply with these prototypes. The user selects the form of equation he desires to use, and completes the equation in accordance with the prototype structure, using the symbols of the particular variables to which the equation is applied. Each equation type defines a single variable on the left-hand side of the equation in terms of some combination of variables on the right-hand side. The equation form is num-bered, and, if the variable it represents is a level or a rate, the number is followed by an L or R, respectively.

It is not always possible to define the model entirely in terms of levels and rates that can be expressed by the given equation forms. It may be necessary or convenient to introduce auxiliary variables. These are defined by using the same equation forms and following the form numbers with the letter A. Initial values may also be entered with the same equation forms by using the letter N. A constant is entered by putting the letter C in column 1 and equating the symbol to its value in the statement field.

Some of the more important equation forms are listed in Table 5-1. The symbol V represents the variable being defined and the symbols P, Q, R, etc. represent other variables. Note that the first two equation forms are marked with an L, indicating that these may only be used to define levels. The others may be used for levels, rates, auxiliary variables, or initial conditions by following the number with L, R, A, or N, respectively.

Table 5-1 DYNAMO Equation Forms

Level Equations	Form Number	Exact Punching Format
$V = V + (DT)(P + Q)$	1L	$V.K = V.J + (DT)(\pm P \pm Q)$
$V = V + (DT)\dfrac{(P + Q)}{Y}$	3L	$V.K = V.J + (DT)(1/\pm Y)(\pm P \pm Q)$
$V = P$	6	$V = \pm P$
$V = P + Q$	7	$V = \pm P \pm Q$
$V = (P)(Q)$	12	$V = (\pm P)(\pm Q)$
$V = (P)(Q + R)$	18	$V = (\pm P)(\pm Q \pm R)$
$V = \dfrac{P}{Q}$	20	$V = \pm P/\pm Q$
$V = \dfrac{P + Q}{R}$	21	$V = (1/\pm Y)(\pm P \pm Q)$
Step Function P at time Q	45	$V = STEP(\pm P, Q)$
$V = MAX(P, Q)$	56	$V = MAX(\pm P, \pm Q)$
$V = MIN(P, Q)$	54	$V = MIN(\pm P, \pm Q)$

As an example of how coding is carried out, consider the flow chart of Fig. 5-2, in which a level INV depends upon a rate INP that increases the level and a rate OUT that decreases the level. Suppose the rates are proportional to two other levels LEV1 and LEV2 with the coefficients of proportionality being 1/T1 and 1/T2, respectively.

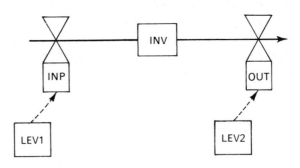

Figure 5-2. Coding of DYNAMO equations.

Alegebraically, the equations can be written

$$INV = INV' + DT(INP - OUT)$$

$$INP = LEV1/T1$$

$$OUT = LEV2/T2$$

The first equation expresses the fact that the present level of INV is the level at the previous interval INV', modified by the changes brought about by the rates during the interval DT. It is a level equation of form 1L. The other two equations are rate equations of form 20. The DYNAMO coding is

1L   INV.K = INV.J + (DT) (INP.JK − OUT.JK)

20R   INP.KL = LEV1. K/T1

20R   OUT.KL = LEV2.K/T2

Equations are coded on FORTRAN forms. The equation form number is punched beginning in column 1, and the equation is punched beginning in column 7, with no intervening blanks. The first blank terminates the equation and the remaining space may be used for comments. Note that every level variable must have the suffix J or K, and every rate variable must have the suffix JK or KL. A rate equation can be coded two ways, depending upon the time period chosen. Thus the equation for INP given above could also have been written:

INP.KL = LEV1.J/T1

The latter form would result in the rate INP being calculated on information that is one unit of time behind the current clock. The evaluation, therefore, is as though there is a delay of one time unit. This is used as a convenient way of introducing a small time lag.

5-5

Delays

The method of representing delays in an Industrial Dynamics model was described in Sec. 4-4. A delay is interpreted as meaning that a rate is proportional to a level. If an output rate called OUT is being controlled by a level called LEV, and the average delay is called DEL, then the delay is written

OUT.KL = LEV.K/DEL.

This is a DYNAMO equation of type 20. A second-order delay can be made by allowing the rate OUT to increase another level LEV1, say, and then decreasing LEV1 by an equation of the form just given. If the second delay is also DEL the over-all delay is $2 \times$ DEL. The process can be repeated indefinitely.

The purpose of using multiple delays is to achieve a response that is more representative of the actual situation being modeled. However, first-order delays are adequate for most purposes.

5-6

Durable Goods Industry Model

To illustrate the way DYNAMO is used in the study of industrial systems, we will construct a model of the business and industrial elements involved in the supply and demand of durable goods. The model is hypothetical and simple; nevertheless, it contains in a rudimentary form the major elements involved in the industry and will serve to demonstrate how Industrial Dynamics models are constructed and used. The example is based on one given originally by Jarmain (2) as an exercise in hand simulation. The present version was written by the staff of the IBM Education Center, Poughkeepsie.

The durable goods industry is considered to be broken into a number of sectors, as illustrated in Fig. 5-3. One sector represents the customers who buy the goods. Orders placed by the customers ultimately affect the other sectors that are represented, namely, the retailer, the distributor, the wholesaler, the manufacturer, and the factory that makes the goods. The retailer supplies customers if he has stock on hand and, to maintain his stock, he places orders with a distributor. The distributor in turn supplies goods to the retailer and places orders on the wholesaler, who repeats the process with the manufacturer. Finally, the factory manufactures new goods to supply the manufacturer.

Each sector orders goods from the next sector. The order rate becomes a demand rate for the next sector. A delay of one time period will be introduced by making the demand rate for each sector equal to the order rate of the preceding sector. For example, the orders placed by the customers are denoted by a rate QORDC (Quantity Ordered by Customers). The resultant input to the retailer is a demand rate QDEMR (Quantity Demanded from Retail). The two rates are equated as follows:

$$QDEMR.KL = QORDC.JK$$

The different time intervals used in the equation introduce a delay of one time period. The shipments from retailer to the customer are represented by a rate QASHR (Quantity Actually Shipped from Retail). As will be explained later,

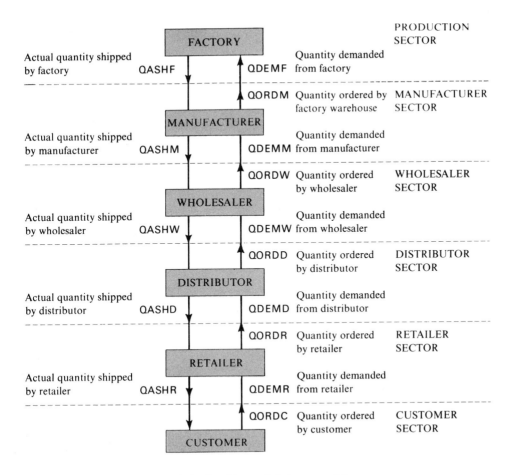

Figure 5-3. Durable goods industry model.

the actual quantity shipped will be a choice between two possible shipping rates. One rate, which will be referred to as Quantity Shipped to Backlog at Retail, is denoted by the symbol QSHBR; the other, to be called Quantity Shipped to Inventory at Retail, is denoted by the symbol QSHIR.

Two levels will be defined for the retail sector. One represents the level of inventory and is called LINVR (Level of Inventory at Retail); the other represents the backlog of orders and is called LBKGR (Level of Backlog at Retail).

Special cases will be made of the customer and factory production sectors. The sectors for the distributor, wholesaler, and manufacturer will be constructed in the same way as the retail sector, using the letters R, D, W, and M to distinguish the retailer, distributor, wholesaler, and manufacturer sectors, respectively. The symbols to be employed for the various quantities that have been defined in the four similar sectors are shown in Table 5-2.

Table 5-2 Symbols Used in DYNAMO Model of Durable Goods Industry

Quantity	Retail	Distribution	Wholesale	Manufacturer
Demand In	QDEMR	QDEMD	QDEMW	QDEMM
Orders Out	QORDR	QORDD	QORDW	QORDM
Shipping to Backlog	QSHBR	QSHBD	QSHBW	QSHBM
Shipping to Inventory	QSHIR	QSHID	QSHIW	QSHIM
Actual Shipping	QASHR	QASHD	QASHW	QASHM
Inventory	LINVR	LINVD	LINVW	LINVM
Backlog	LBKGR	LBKGD	LBKGW	LBKGM

It will be noticed that all rates are designated by a symbol starting with Q and all levels are designated with a symbol starting with L. This is a matter of convenience; no particular symbolism is required by the DYNAMO program.

The time interval DT to be used in the model will be set to 1, and it can conveniently be thought of representing 1 day.

5-7

Customer Sector

The prime objective of an Industrial Dynamics study is to discover how an industry can be expected to respond to demands. The model will be initialized to represent the steady state conditions resulting from a period of constant demand, and the system response will be tested by introducing a sudden increase of demand to a new level. One of the equation forms available in DYNAMO, number 45, is a step function which indicates the level to which a variable is to be increased and the time at which the increase occurs. The following statements will be used to initialize the demand to 4 items/day and increase the demand to 6 items/day after 4 days:

$$6N \qquad QORDC = 4$$

$$45R \qquad QORDC.KL = STEP(6,4)$$

5-8

Retail Sector

The sector concerned with the retailer's operations will be described next, and, as will be seen, it essentially repeats the model of an inventory control system given in Sec. 4-9. A block diagram of the retail sector is given in Fig. 5-4. Entering the sector from the customer sector is a flow of orders, controlled by

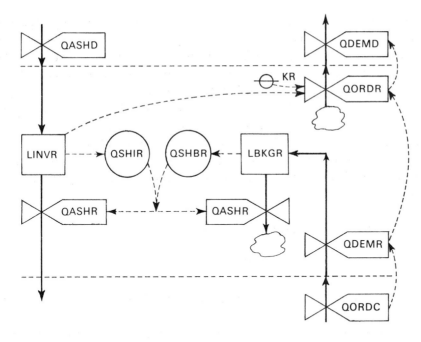

Figure 5-4. Retail section.

the rate QORDC. This becomes a demand rate QDEMR delayed one day by using the equation

$$6R \qquad QDEMR.KL = QORDC.JK \qquad (5\text{-}1)$$

Similarly, the retailer sends orders to the distributor at the rate QORDR which becomes the demand rate QDEMD for the distributor.

Goods are received by the retailer from the distributor at a rate QASHD and they are delivered to customers at the rate QASHR. These two rates, therefore, control the inventory level. The order backlog level is also diminished by the rate QASHR as deliveries are made; it is increased by the rate QDEMR. Equations can immediately be written for the two levels. These are

$$1L \qquad LBKGR.K = LBKGR.J + (DT)(QDEMR.JK - QASHR.JK) \qquad (5\text{-}2)$$

$$1L \qquad LINVR.K = LINVR.J + (DT)(QASHD.JK - QASHR.JK) \qquad (5\text{-}3)$$

A level in DYNAMO can be positive or negative, and there is no direct way of preventing the level from changing sign. It is not realistic for the inventory level to go negative and the two quantities QSHBR (Quantity Shipped to Backlog at Retail) and QSHIR (Quantity Shipped to Inventory at Retail) have been defined for the purpose of preventing a negative inventory. There is assumed to be a one-day delay in delivering goods. If there is enough stock in the inventory to cover the backlog of orders, the rate of shipping will be QSHBR set to reduce the backlog at a rate proportional to the backlog level. If there is not sufficient inventory on hand, the rate will be set to deliver as much stock as there is in the inventory, and the rate of delivery for this condition is QSHIR. The choice is to select the smaller of these two rates. So long as the inventory exceeds the backlog, shipping will occur at the backlog rate; but when the inventory is too low, shipping will occur at the inventory rate. A special equation form allows this choice to be made. The shipping rates are defined as follows:

$$20A \qquad QSHBR.K = LBKGR.K/DT \tag{5-4}$$

$$20A \qquad QSHIR.K = LINVR.K/DT \tag{5-5}$$

$$54R \qquad QASHR.KL = MIN(QSHBR.K, QSHIR.K) \tag{5-6}$$

To complete the retail sector, the ordering policy must be established. The policy that was tested in Sec. 4-9 will be used again. The order rate will be the demand rate plus a term proportional to the difference between the actual inventory level and a desired level.

A constant called LINDR (Level of Inventory Desired at Retail) is defined and the symbol KR, called the ordering constant, is used to denote the coefficient of proportionality. For convenience, the auxiliary variable LDEVR (Level of Deviation at Retail) is also introduced. The ordering policy is then described by the equations:

$$18A \qquad LDEVR.K = (KR)(LINDR - LINVR.K) \tag{5-7}$$

$$7A \qquad QORDR.KL = QDEMR.JK + LDEVR.K \tag{5-8}$$

The set of Eqs. (5-1) to (5-8) now define the retail sector completely.

5-9

Other Sectors

The sectors concerned with the distributor, wholesaler, and manufacturer will be modeled in the same way as the retail sector, except that different symbols will be used and the ordering policy will be simpler. The symbols used are those given in Table 5-2. The ordering rate for each sector will simply be equated to the demand rate with a delay of one day. For example, in the distribution sector, the ordering is described by the equation

$$6R \qquad QORDD.KL = QDEMD.JK$$

The operation of the factory will be represented by three successive work stages, each resulting in a one-day delay. Letting WIP1, 2, and 3 represent the outputs of the stages, the equations representing the factory are

$$6R \qquad QDEMF.KL = QORDM.JK$$

$$6R \qquad WIP1.KL = QDEMF.JK$$

$$6R \qquad WIP2.KL = WIP1.JK$$

$$6R \qquad WIP3.KL = WIP2.JK$$

$$6R \qquad QASHF.KL = WIP3.JK$$

5-10
Initial Conditions

As was done in Sec. 4-9, the system will be tested under the condition that it was initially supplying customers with a steady rate of 4 items/day, and that all inventory levels were stabilized at 20 items, which is the desired level. After four days, the demand will suddenly increase to 6 items/day. The two equations given in Sec. 5-7 defined the customer demand pattern. To establish the initial levels and rates for the retail section, the following equations are required:

$$6N \qquad LINVR = 20$$

$$6N \qquad LBKGR = 4$$

$$6N \qquad LDEVR = 0$$

$$C \qquad LINDR = 20$$

$$6N \qquad QDEMR = 4$$

$$6N \qquad QORDR = 4$$

$$6N \qquad QASHR = 4$$

$$C \qquad KR = 0.5$$

The last initial condition establishes an ordering constant of $\frac{1}{2}$.

For the other sectors, the inventory levels will be set to 20 and all rates to 4 items/day.

5-11

Coding of the Problem

Coding for the complete model is shown in Fig. 5-5. It will be noticed that various lines of comment and blank spaces have been introduced into the listing. These are inserted by using a card having the word NOTE in columns 1–4. The first card in the listing carries an asterisk in column 1, and is a control card indicating that the program is a DYNAMO run and the run is expected to produce results (as opposed to being a TEST compilation). The second card is a RUN card that provides the program with an identification label that is printed along with the page number on every page of output. The model itself follows. Finally, there are two more control cards at the end. The card marked SPEC is a specification card that gives the size of DT and the length of the run in time units. The last two items on the card and the final card relate to program output.

In this program, the ordering constant has been inserted directly in the model rather than as an initial constant. This requires equation type 21 to be used in the retail section instead of type 18 as shown in Sec. 5-8.

5-12

DYNAMO Output

It is possible both to print tables and plot graphs as part of the DYNAMO output. The last two fields on the SPEC card indicate the intervals of time at which to print or plot output respectively. Here, the request is for no tables but a plot at every time unit.

The plot output of DYNAMO is a particularly useful feature of the program. Up to 10 variables can be plotted. The choices are named and identified by a single character on the last control card. The program will type a line of print at every tabulation interval, with the time axis running along the length of the paper. The values of the variables are indicated by printing their identification character at the appropriate distance across the width of the paper. The program automatically scales the variables and includes in the print-out scale graduations along both axes.

Figure 5-6 shows the output for the model that has just been described. The variables plotted and their identification characters are described across the top. Where two variables take the same value at one time interval, the program prints one character and makes a note on the extreme right to indicate which other characters take the same position.

```
*        12345,DYN,RESULT,1,1,0,0
RUN      DYNAMO SIMULATION OF DURABLE GOODS INDUSTRY
NOTE
NOTE              CUSTOMER SECTOR
NOTE
  6N     QORDC=4                               INITIAL ORDER RATE
 45R     CORDC.KL=STEP(6,4)                    ORDER RATE INCREASE
NOTE
NOTE              RETAIL SECTOR
NOTE
  6R     QDEMR.KL=QORDC.JK                     ORDERS RECEIVED
  1L     LBKGR.K=LBKGR.J+(DT)(QDEMR.JK-QASHR.JK)   ORDER BACKLOG LEVEL
  1L     LINVR.K=LINVR.J+(DT)(QASHD.JK-QASHR.JK)   INVENTORY LEVEL
 20A     QSHBR.K=LBKGR.K/DT                    SHIP. RATE TO BACKLOG
 20A     QSHIR.K=LINVR.K/DT                    SHIP. RATE TO INVENTORY
 54R     CASHR.KL=MIN(QSHBR.K,QSHIR.K)         ACTUAL SHIP. RATE
 21A     LDEVR.K =(1/2)(LINVR-LINVR.K)         ADJUSTMENT TO ORDER RATE
  7A     QORDR.KL=QDEMR.JK+LDEVR.K             ORDER RATE
NOTE
NOTE     INITIAL CONDITIONS
NOTE
  6N     LINVR=20                              INITIAL INVENTORY
  6N     LBKGR=4                               INITIAL BACKLOG
  6N     QDEMR=4                               INITIAL DEMAND RATE
  6N     QORDR=4                               INITIAL ORDER RATE
  6N     QASHR=4                               INITIAL SHIP. RATE
  6N     LDEVR=0                               INITIAL ORDER ADJUSTMENT
  C      LINDR=20                              DESIRED INVENTORY LEVEL
NOTE
NOTE              DISTRIBUTOR SECTOR
NOTE
  6R     QDEMD.KL=QORDR.JK                     ORDERS RECEIVED
  1L     LBKGD.K=LBKGD.J+(DT)(QDEMD.JK-QASHD.JK)   ORDER BACKLOG LEVEL
  1L     LINVD.K=LINVD.J+(DT)(QASHW.JK-QASHD.JK)   INVENTORY LEVEL
 20A     QSHBD.K=LBKGD.K/DT                    SHIP. RATE TO BACKLOG
 20A     QSHID.K=LINVD.K/DT                    SHIP. RATE TO INVENTORY
 54R     QASHD.KL=MIN(QSHBD.K,QSHID.K)         ACTUAL SHIP. RATE
  6R     QORDD.KL=QDEMD.JK                     ORDER RATE
NOTE
NOTE     INITIAL CONDITIONS
NOTE
  6N     LINVD=20                              INITIAL INVENTORY
  6N     LBKGD=4                               INITIAL BACKLOG
  6N     QDEMD=4                               INITIAL DEMAND RATE
  6N     CORDD=4                               INITIAL ORDER RATE
  6N     QASHD=4                               INITIAL SHIP. RATE
NOTE
NOTE              WHOLESALE SECTOR
NOTE
  6R     QDEMW.KL=QORDD.JK                     ORDERS RECEIVED
  1L     LBKGW.K=LBKGW.J+(DT)(QDEMW.JK-QASHW.JK)   ORDER BACKLOG LEVEL
  1L     LINVW.K=LINVW.J+(DT)(QASHM.JK-QASHW.JK)   INVENTORY LEVEL
 20A     QSHBW.K=LBKGW.K/DT                    SHIP. RATE TO BACKLOG
 20A     QSHIW.K=LINVW.K/DT                    SHIP. RATE TO INVENTORY
 54R     QASHW.KL=MIN(QSHBW.K,QSHIW.K)         ACTUAL SHIP. RATE
  6R     CORDW.KL=QDEMW.JK                     ORDER RATE
NOTE
NOTE     INITIAL CONDITIONS
NOTE
  6N     LINVW=20                              INITIAL INVENTORY
  6N     LBKGW=4                               INITIAL BACKLOG
  6N     QDEMW=4                               INITIAL DEMAND RATE
  6N     QORDW=4                               INITIAL ORDER RATE
  6N     QASHW=4                               INITIAL SHIP. RATE
NOTE
NOTE              MANUFACTURING SECTOR
NOTE
  6R     QDEMM.KL=CORDW.JK                     ORDERS RECEIVED
  1L     LBKGM.K=LBKGM.J+(DT)(QDEMM.JK-QASHM.JK)   ORDER BACKLOG LEVEL
  1L     LINVM.K=LINVM.J+(DT)(QASHF.JK-QASHM.JK)   INVENTORY LEVEL
 20A     QSHBM.K=LBKGM.K/DT                    SHIP. RATE TO BACKLOG
 20A     QINVM.K=LINVM.K/DT                    SHIP. RATE TO INVENTORY
 54R     QASHM.KL=MIN(QSHBM.K,QSHIM.K)         ACTUAL SHIP. RATE
  6R     QORDM.KL=QDEMM.JK                     ORDER RATE
NOTE
NOTE     INITIAL CONDITIONS
NOTE
  6N     LINVM=20                              INITIAL INVENTORY
  6N     LBKGM=4                               INITIAL BACKLOG
  6N     QDEMM=4                               INITIAL DEMAND RATE
  6N     QORDM=4                               INITIAL ORDER RATE
  6N     QASHM=4                               INITIAL SHIP. RATE
NOTE
NOTE              FACTORY SECTOR
NOTE
  6R     QDEMF.KL=QORDM.JK                     ORDERS RECEIVED
         QPFS1.KL=QDEMF.JK                     WORK STAGE 1
         QPFS2.KL=QPFS1.JK
         QPFS3.KL=QPFS2.JK                     WORK STAGE 3
         QASHF.KL=QPFS3.JK                     ACTUAL SHIP. RATE
NOTE
NOTE     INITIAL CONDITIONS
NOTE
  6N     QDEMF=4
  6N     QPFS1=4
  6N     QPFS2=4
  6N     QPFS3=4
  6N     QASHF=4                               INITIAL SHIP. RATE
NOTE
SPEC     DT=1/LENGTH=50/PRTPER=0/PLTPER=1
PLOT     QORDC=C,LINVR=R,LINVD=D,LINVW=W,LINVM=M
```

Figure 5-5. Coding of durable goods industry model.

QOKDC=C, LINVR=R, LINVD=D, LINVW=W, LINVM=M

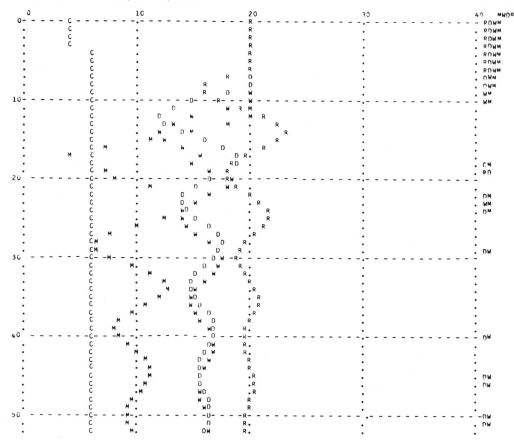

Figure 5-6. Results of DYNAMO run.

The output shows the levels of all the inventories. The inventory control policy of the retailer results in his inventory level returning to its planned level. The other sectors suffer losses because their ordering policies simply equate orders to demand. The delays in transmitting orders are apparent in the fact that the sectors respond to the initial change of demand at different times.

A much more comprehensive DYNAMO model of an industry will be found in Ref. (3). In addition, a DYNAMO model of an inventory control system that includes exponential smoothing of customer demand is given in (1). For other DYNAMO examples, see the bibliography of Ref. (4) of Chap. 4.

Exercises   5-1   Program the durable goods industry model assuming that the distributor, wholesaler, and manufacturer all use the same inventory control policy as the retailer.

5-2 Program in DYNAMO the air conditioner sales model of Sec. 4-4, allowing for breakdowns. Assume the average time to sell a house is 5 months, the average time to install an air conditioner is 10 months, and breakdowns occur, on the average after 25 months. Take the initial housing market to be 1,000 houses.

5-3 Program problem 4-5 in DYNAMO.

5-4 A company establishes a pension fund for its employees by setting aside $P a month for each employee. The accumulated fund is invested and earns 5% per annum. The work force is expected to grow at 3% per annum. The company wants to study the soundness of its plan and proposes to simulate the effects of different assumptions about average length of service and average length of retirement. Draw an Industrial Dynamics flow chart for the simulation.

5-5 The birth rate of a country is adding 100,000 people a year to an initial population of 5,000,000. The average life expectancy is 65 years. It is estimated that one ton of coal is consumed per annum for each individual. Draw a diagram showing how the country's resources of 500,000,000 tons are being depleted.

5-6 Two competing companies invest funds in capital equipment to improve their positions. The rate at which each invests funds decreases linearly as their own investment increases but increases linearly as their competitor's investment increases. Draw a diagram from which to simulate the competition and determine under what conditions the investments will stabilize.

5-7 A television rental company installs and maintains sets. It has a work force that is trained to both install and repair sets. Sets are installed and repaired at a rate proportional to the number of men assigned respectively to installing and repairing. Once installed, sets break down after a known average lifetime. The employer tries to keep his men on installation work but, to balance the workload, he transfers men to repair work at a rate proportional to the number of outstanding bad sets. The repairs take a certain average time, following which the men return to installation work. Draw a diagram and write a DYNAMO program to simulate the business.

Bibliography   1   Pugh, Alexander L., III, *DYNAMO User's Manual*, 2nd Ed., Cambridge, Mass.; The M.I.T. Press, 1963.

2   Jarmain, W. Edwin (ed.), *Problems in Industrial Dynamics*, Cambridge, Mass: The M.I.T. Press, 1963.

3   Forrester, Jay W., *Industrial Dynamics*, Cambridge, Mass.: The M.I.T. Press, 1961.

6

PROBABILITY
CONCEPTS
IN SIMULATION

6-1

**Stochastic
Variables**

Previous chapters have discussed models which are deterministic; that is, the exact outcome of any activity can be described and the precise history of the system can be traced by following the effects of the activities. In practice, many systems include activities that behave randomly, in which case the exact sequence of events is not known. Sometimes the activity is not intrinsically random in nature, but complete information about its behavior is not known and it must be described as a random activity. Possibly, complete information is available or can be derived, but it is so complex that it is more convenient to describe the activity as being random.

A variable representing the outcome of a random activity is said to be a *stochastic variable*. Although the exact sequence of values taken by a stochastic variable is not known, the range of values over which it can vary, and the probability with which it will take the values, may be known or assumed to be known. Stochastic variables are therefore discussed in terms of functions which describe the probability of the variable taking various values.

Because of the interrelations between the activities of a system, the introduction of a stochastic variable into a system becomes reflected throughout the system. Most, if not all, of the quantities of interest in measuring the system performance then show random fluctuations. A simulation study then includes the task of deriving the probability functions of some system variables, together with their expected values and standard deviations.

6-2

Probability Functions

If a stochastic variable can take n different values x_i $(i = 1, 2, 3, \ldots, n)$ and the probability of the value x_i being taken is p_i, the set of numbers p_i $(i = 1, 2, 3, \ldots, n)$ is said to be a *discrete probability function*. Since the variable must always take one of the values p_i, it follows that $\sum_{i=1}^{n} p_i = 1$. Typical examples of discrete variables that can occur in a simulation study are the number of items a customer buys in a store or the origin of a message in a communication system where the terminals creating messages have been numbered from 1 to n.

The distribution may be a known set of numbers. For example, with a die, the probability of each of the six faces is $\frac{1}{6}$ but, frequently, a discrete probability function must be estimated by counting how many times each possible value occurs in a sample. The probability is then estimated by computing the fraction of all observations that took each value. Table 6-1, for example, represents data gathered on the number of items customers bought in a supermarket for a sample of 250 customers. The third column is the estimate of the probability of buying n items, derived by dividing the number of customers who bought n items by the total number of observations.

Table 6-1 Number of Items Bought by Customers

No. of Items	No. of Customers	$p(n)$
1	25	0.10
2	128	0.51
3	47	0.19
4	38	0.15
5	12	0.05
	250	1.00

When the process being observed is continuous and not limited to discrete values an infinite number of possible values can be assumed by the variable. The probability of any one specific value occurring must logically be considered to be zero. To describe the variable, a *probability density function*, $f(x)$, is defined. The probability that x falls in the range x_1 to x_2 is given by

$$\int_{x_1}^{x_2} f(x)dx \qquad f(x) \geqslant 0$$

Intrinsic in the definition of a probability density function is the property that the integral of the probability density function taken over all possible values is 1. That is,

$$\int_{-\infty}^{\infty} f(x)dx = 1$$

A related function is the *cumulative distribution function* which defines the probability that the observed value is less than or equal to x. If $F(x)$ is the cumulative probability distribution function, then

$$F(x) = \int_{-\infty}^{x} f(x)dx$$

From its definition, $F(x)$ is a positive number ranging from 0 to 1, and the probability of x falling in the range x_1 to x_2 is $F(x_2) - F(x_1)$. Figure 6-1 illustrates a probability density function and the corresponding cumulative distribution function.

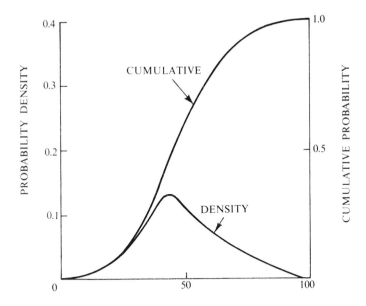

Figure 6-1. Probability distribution functions.

The mean, or expected value, of a distribution is given by

$$\mu = \int_{-\infty}^{\infty} xf(x)dx$$

In the case of a discrete distribution, the mean is evaluated from the formula

$$\mu = \sum x_i p(x_i)$$

where the sum is taken over all values of x at which the distribution is defined.

6-3

Numerical Derivation of Continuous Probability Functions

As with a discrete probability function, it is possible that a mathematical formula can be derived or assumed for a continuous probability density function and hence for the cumulative distribution function. Frequently, however, the function must be numerically estimated from a series of observations. The observations will count how often the value of a variable falls within certain intervals, and the probability function must be estimated from these observations.

The customary way of organizing data derived from observations is to display them as a *frequency distribution*, which shows the number of times the variable falls in different intervals. For example, suppose that 1,000 observations of telephone call lengths are made and tabulated in intervals of 10 seconds. The frequency distribution might then appear as shown in Table 6-2. The leftmost column defines a number of intervals, and the next column records the number of calls whose length fell within that interval. The distribution can also be displayed graphically in a bar-graph, as shown in Fig. 6-2.

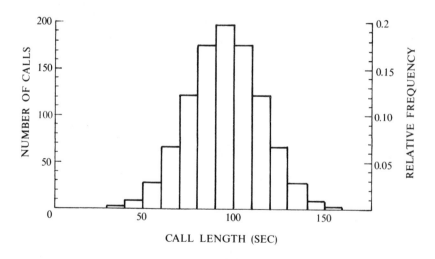

Figure 6-2. Frequency distribution of telephone call lengths.

Table 6-2   Distribution of Telephone Call Lengths

Call Length (sec)	Number of Calls	Relative Frequency	Probability Density	Cumulative Distribution
0	0	0.000	0.0000	0.000
10	1	0.001	0.0001	0.001
20	1	0.001	0.0001	0.002
30	1	0.001	0.0001	0.003
40	2	0.002	0.0002	0.005
50	8	0.008	0.0008	0.013
60	28	0.028	0.0028	0.041
70	65	0.065	0.0065	0.106
80	121	0.121	0.0121	0.227
90	175	0.175	0.0175	0.402
100	197	0.197	0.0197	0.599
110	175	0.175	0.0175	0.774
120	121	0.121	0.0121	0.895
130	65	0.065	0.0065	0.960
140	28	0.028	0.0028	0.988
150	9	0.009	0.0009	0.997
160	2	0.002	0.0002	0.999
170	1	0.001	0.0001	1.000

A more useful way of describing the same information is as a *relative fre-quency distribution*, in which the number of observations for each interval is divided by the total number of observations. Table 6-2 shows the relative fre-quency in the third column and Fig. 6-2 displays the relative frequency on the axis to the right of the figure.

From the definition, it is apparent that the sum of all the values of the relative frequency distribution is 1. It is important to note, however, that the relative frequency distribution is *not* necessarily the probability density function for the variable. If the n intervals for which the frequency distribution has been gener-ated are $x_i < x \leqslant x_{i+1}$ ($i = 0, 1, \ldots, n - 1$), and p_i is the relative frequency count for the ith interval, then p_i must be interpreted as the integral of the probability density function over the interval; that is,

$$p_i = \int_{x_i}^{x_{i+1}} f(x)dx$$

The choice of the interval size will normally be made on the assumption that it is small enough to ignore variations of the probability density function over the interval. If so, the value of the density function in each interval is $p_i/(x_{i+1} - x_i)$. For the case of the data in Table 6-2, for example, the relative frequency distribution for each interval is divided by 10 to get the approximation to the probability density function shown in the table as the fourth column. If the intervals for which the observations were taken are not all the same size, appropriate adjustments must be made.

In practice, there is more interest in deriving the cumulative distribution function. If the values of p_i are accumulated, the successive values, $F_r = \sum_{i=1}^{r} p_i$ $(r = 1, 2, \ldots, n)$, represent the value of the cumulative distribution function at the points x_r $(r = 1, 2, \ldots, n)$. This is shown as the fifth column of Table 6-2. A series of straight line segments drawn through these points, as illustrated in Fig. 6-3, can be taken as an approximation to the cumulative distribution function. The straight line approximation conforms with the previous assumption that the probability density function is constant over each interval. A better approximation can be attempted by fitting a smooth curve to the points; in which case, the probability density function can be derived by plotting the slope of the cumulative distribution curve.

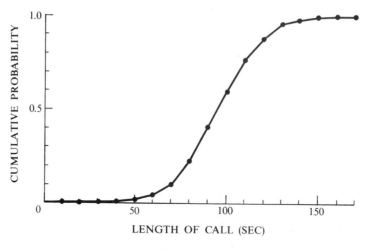

Figure 6-3.

6-4

**Monte Carlo
Simulation**

Probability distributions have been discussed so far as a means of describing the behavior of a stochastic variable. They record the results of a number of observations, considering only the values assumed by the variable and not the sequence in which they occurred. In carrying out a simulation that involves stochastic variables, the reverse problem arises. It is necessary to generate a

Probability Concepts in Simulation   /   Ch. 6

sequence of numbers where the successive values are random and have the distribution that describes the stochastic variable.

Various devices and techniques exist for producing such sequences. The simplest example is for a discrete probability distribution function where the choice is between n different numbers, each occurring with equal probability. A roulette wheel that has the same number of sections as there are different values, or a die with that number of faces will generate the sequence.

Because of this analogy, the name Monte Carlo with its connection with gambling has become a general term used to describe any computational method using random numbers. *Monte Carlo methods* are used in a wide variety of problems such as the evaluation of complicated integrals, studies of population growth, and the design of nuclear reactors (1); such cases are often described as *Monte Carlo simulations*. The term system simulation was defined in Chap. 2 to mean a numerical calculation in which successive states of a system are determined over time.

Many applications described as Monte Carlo simulations are simulations in this sense, because they follow the progress of a stochastic process. Some Monte Carlo methods, however, are applied to problems which are not intrinsically random; the random numbers merely provide a convenient way to evaluate a quantity; for example, the value of an integral. An integral can be interpreted as representing an area or volume. The Monte Carlo method generates at random the coordinates of a point within a space and tests whether the point falls within the area or volume defined by the integral. By repeating the experiment many times and measuring the proportion of points that fall within the area or volume, an approximate value for the integral is derived.

Hammersley and Handscomb (1) distinguish between two types of Monte Carlo methods by defining those that apply to problems in which a random process is intrinsic as probabilistic methods, and the others as deterministic. It is the probabilistic type of application that is used in system simulation.

A Monte Carlo simulation requires sequences of random numbers which are drawn from a distribution that, in general, is not uniform. Methods for directly generating random numbers with a particular distribution are not usually available. Methods do, however, exist for generating random numbers with a uniform distribution. Almost all methods used for generating non-uniform distributions are based on the principle of transforming a uniformly distributed sequence of random numbers into the required sequence.

6-5

Continuous Uniformly Distributed Random Numbers

By a continuous uniform distribution we mean that the probability of a variable x falling in any interval within a certain range of values is proportional to the ratio of the interval size to the range; that is, every point in the range is equally likely to be chosen. Suppose the possible range of values is from A to B $(B > A)$, then the probability that x will fall in an interval Δx is $\Delta x/(B - A)$. Drawn as a graph, the probability density function is a straight line of height $1/(B - A)$ between the points A and B.

There is no loss in generality in assuming that the range is from 0 to 1, because, if x_i is a sequence of uniformly distributed numbers in the range 0 to 1, then $(B - A)x_i + A$ is a uniformly distributed sequence in the range A to B. The following discussion will therefore consider the case of the range being from 0 to 1, so that the probability density function is

$$f(x) = 1 \qquad 0 \leqslant x \leqslant 1$$
$$= 0 \qquad \text{elsewhere}$$

In dealing with numerical calculations, a certain number of digits are chosen to represent a quantity, depending upon the desired accuracy. Strictly speaking, it is not then possible to represent a continuous variable, since the finite number of digits allows only a finite number of possible values. Given enough digits, the number of possible values that can be assumed is sufficiently large to treat the variable as being continuous. It will be assumed that the generated random numbers are represented by sufficient digits to make the "granularity" caused by the finite number of digits negligible. Caution must be used, however, if a random number is modified subsequently to its generation. If the number is multiplied by a factor that is greater than 1, there is an increase in "granularity" which can lead to a loss of accuracy. Most digital computer programs using a random number generator use a full computer word to represent the random number, so that the number has sufficient binary digits. However, for accurate calculations, it is not unusual to use double precision for the calculation of random numbers.

There are many physical processes that can be considered to give a sequence of uniformly distributed random numbers. For example, an electrical pulse generator can be made to drive a counter cycling between 0 and 9. By sampling the counter at random intervals, the value can be taken as a digit of the desired random number. By repeating the process many times, or running several counters in parallel, a random number of any desired number of digits can be created.

Tables of uniformly distributed random numbers produced by such physical processes have been published. Reference (2) is a particularly comprehensive set of numbers generated by a method based on the process just described. Table 6-3 is a set of random numbers reproduced from a single page of Ref. (2). The digits have been arranged in columns of five, and the columns and rows have been further divided into groups for ease of reading.

Suppose it is decided to generate a series of uniformly distributed random numbers between 0 and 1 to an accuracy of 1 part in 10,000; that is, the numbers are to have 4 decimal digits. The first column on the left can be read row-by-row discarding, say, the last digit, and when the column is finished the next column can be used. Using this method, we find the first five random numbers are

$$0.1009$$
$$0.3754$$
$$0.0842$$
$$0.9901$$
$$0.1280$$

If numbers with more than five digits are needed, columns are combined. It is not necessary to start with the first column. In fact, for proper use, particularly when a calculation is being repeated with different sets of random numbers, the starting points should be chosen at random. This could be done by using a random number to decide upon the page, column, and row at which to start.

6-6 Computer Generation of Random Numbers

It is possible to read tables of numbers such as Table 6-3 into a computer, but the required quantity of random numbers can be very large and it is very inconvenient to prepare data in this form. As a result, computer programs have been developed for generating sequences of numbers uniformly distributed over a given range (usually 0 to 1). Starting with an initial number, the methods most commonly used specify a procedure of generating a second number. Using the second number as input, the procedure is repeated to produce a third number, and so on. In general, the $(i + 1)$th number of the sequence is generated from the ith number. An extensive literature exists on the generation of random numbers. (See Ref. (3) and its bibliography.)

The particular method that has come to be most widely used is called the congruence method or, sometimes, the residue method. Given three constants λ, μ, and P, the procedure derives the $(i + 1)$th number from the ith number by multiplying by λ, adding μ, and then taking the remainder, or residue, upon dividing by P; this procedure is described mathematically by the expression

$$c_{i+1} = (\lambda c_i + \mu) \quad (\text{modulo } P)$$

To begin the process, an initial number c_0 is needed, and this is called a seed. Complete specification of the process therefore needs four numbers, λ, μ, c_0, and P.

Table 6-3   Table of Random Digits

10097	32533	76520	13586	34673	54876	80959	09117	39292	74945
37542	04805	64894	74296	24805	24037	20636	10402	00822	91655
08422	68953	19645	09303	23209	02560	15953	34764	35080	33606
99019	02529	09376	70715	38311	31165	88676	74397	04436	27659
12807	99970	80157	36147	64032	36653	98951	16877	12171	76833
66065	74717	34072	76850	36697	36170	65813	39885	11199	29170
31060	10805	45571	82406	35303	42614	86799	07439	23403	09732
85269	77602	02051	65692	68665	74818	73053	85247	18623	88579
63573	32135	05325	47048	90553	57548	28468	28709	83491	25624
73796	45753	03529	64778	35808	34282	60935	20344	35273	88435
98520	17767	14905	68607	22109	40558	60970	93433	50500	73998
11805	05431	39808	27732	50725	68248	29405	24201	52775	67851
83452	99634	06288	98083	13746	70078	18475	40610	68711	77817
88685	40200	86507	58401	36766	67951	90364	76493	29609	11062
99594	67348	87517	64969	91826	08928	93785	61368	23478	34113
65481	17674	17468	50950	58047	76974	73039	57186	40218	16544
80124	35635	17727	08015	45318	22374	21115	78253	14385	53763
74350	99817	77402	77214	43236	00210	45521	64237	96286	02655
69916	26803	66252	29148	36936	87203	76621	13990	94400	56418
09893	20505	14225	68514	46427	56788	96297	78822	54382	14598
91499	14523	68479	27686	46162	83554	94750	89923	37089	20048
80336	94598	26940	36858	70297	34135	53140	33340	42050	82341
44104	81949	85157	47954	32979	26575	57600	40881	22222	06413
12550	73742	11100	02040	12860	74697	96644	89439	28707	25815
63606	49329	16505	34484	20419	52563	43651	77082	07207	31790
61196	90446	26457	47774	51924	33729	65394	59593	42582	60527
15474	45266	95270	79953	59367	83848	82396	10118	33211	59466
94557	28573	67897	54387	54622	44431	91190	42592	92927	45973
42481	16213	97344	08721	16868	48767	03071	12059	25701	46670
23523	78317	73208	89837	68935		26252	29663	05522	82562
04493	52494	75248	33824	45862	51025	61962	79335	65337	12472
00549	97654	64051	88159	96119	63896	54692	82391	23287	29529
35963	15307	26898	09354	33351	35462	77974	50024	90103	39333
59808	08391	45427	26842	83609	49700	13021	24892	78565	20106
46058	85236	01390	92286	77281	44077	93910	83647	70617	42941
32179	00597	87379	25241	05567	07007	86743	17157	85394	11838
69234	61406	20117	45204	15956	60000	18743	92423	97118	96338
19565	41430	01758	75379	40419	21585	66674	36806	84962	85207
45155	14938	19476	07246	43667	94543	59047	90033	20826	69541
94864	31994	36168	10851	34888	81553	01540	35456	05014	51176
98086	24826	45240	28404	44999	08896	39094	73407	35441	31880
33185	16232	41941	50949	89435	48481	88695	41994	37548	73043
80951	00406	96382	70774	20151	23387	25016	25298	94624	61171
79752	49140	71961	28296	69861	02591	74852	20539	00387	59579
18633	32537	98145	06571	31010	24674	05455	61427	77938	91936
74029	43902	77557	32270	97790	17119	52527	58021	80814	51748
54178	45611	80993	37143	05335	12969	56127	19255	26040	90324
11664	49883	52079	84827	59381	71539	09973	33440	88461	23356
48324	77928	31249	64710	12295	36870	32307	57546	15020	09994
69074	94138	87637	91976	35584	04401	10518	21615	01848	76938

The number P is usually taken to be a power of the number system being used, so that, for decimal numbers, P is of the form 10^n. Consider the numbers c_i, λ, and μ as being decimal integers. Then the quantity $\lambda c_i + \mu$ is an integer. Dividing by P, when it has the form 10^n, is simply the process of moving the decimal point n places to the left. The remainder c_{i+1} is then an n digit decimal fraction between 0 and 1, and it is taken to be the desired random number. Dropping the decimal point and taking c_{i+1} as input, the process is repeated to derive the next random number. Most digital computers use a binary number system so the number P is of the form 2^n, but the same procedure applies to produce a series of binary random numbers.

The product of two numbers of n digits is a number of $2n$ digits. Digital computers normally use a fixed word size to represent numbers, and a double length word to accommodate the double length product that results from a multiplication. Alternatively, they allow overflow to occur and retain the lower n digits. By choosing P to be the number 2^n where n is the number of digits in the computer fixed word size, the process of dividing and extracting the residue is the simple process of accepting the lower half of the product or the residue left after overflow. It is possible to choose P for convenience in this manner but there are restrictions on the choice of the numbers λ, μ, and c_0. For the produced numbers to be random, the digits of the numbers must be individually random and uncorrelated with each other. By carefully choosing the constants λ, μ, and the seed c_0, it is possible to meet these conditions to a high degree of significance.

It is apparent that it is impossible to produce a nonrepeating sequence of numbers from the procedure. The set of numbers represented by n digits is finite, so ultimately some member of the sequence repeats an earlier value and the sequence is repeated thereafter. However, careful selection of the constants results in a very long sequence before such a repetition occurs (4), so that for practical purposes the sequence can be said to produce random numbers. Because of the number repetitiveness of the sequence, the term *pseudo-random number generator* is used to describe the procedure. Henceforth, the unqualified use of the term random number will mean a uniformly distributed random number drawn from a pseudo-random number generator.

Three types of congruence pseudo-random number generators have been used. They are additive, multiplicative, and mixed congruential generators. A mixed method is defined by the formula already discussed; namely

$$c_{i+1} = (\lambda c_i + \mu) \quad \text{(modulo } P)$$

If $\lambda = 1$, the method is said to be additive, and, if $\mu = 0$, the method is said to be multiplicative. Recent research (5) suggests that the multiplicative method is superior to the additive method and that mixed methods are not noticeably better than simple multiplicative methods.

The problem of choosing the correct values for the constants of a pseudo-random number generator is a complex one. In particular, the choice of good values depends upon the word size of the computer being used. The need for random numbers in computations is such that good pseudo-random number generators have been created and tested for most computers, and users of digital computers usually have access to a well-tested random number generator.

6-7

A Uniform Random Number Generator

Although the number of computational steps involved in a congruence method of generating a random number is small, so that no great penalty is paid for repeating the coding wherever a random number is needed, it is customary to create a subroutine that can be used as a common source of all random numbers.

There is often a need for both floating-point and integer random numbers, and generators are sometimes written to produce both forms of number. The following is an example designed for a 32 bit word size. It employs a simple multiplicative congruence method with $\lambda = 5^{13}$ (1,220,703,125), Examples of generators for other word sizes are given in Ref. (5).

```
        SUBROUTINE RAND (IX,IY,YFL)

        IY = IX★1220703125

        IF (IY) 1,2,2

1       IY = IY + 214783647 + 1

2       YFL = IY

        YFL = YFL★0.4656613E-9

        RETURN

        END
```

The program is executed by issuing the statement:

```
        CALL RAND(IX,IY,YFL)
```

The outputs are IY, a random integer between 1 and $2^{31} - 1$, and YFL, a floating-point random number between 0 and 1. The main program that calls the generator must provide as a seed, an initial value of IX in the form of an odd integer of less than nine decimal digits. Upon each execution of the subroutine, the output IY must be substituted for the input IX, in preparation for the next execution. Any number of independent random number sequences can be derived from one copy of the subroutine by defining different parameters, IX1, IX2, ..., one for each sequence. If only one sequence is needed, the substitution of the output for the next input can be performed within the subroutine by including the statement IX = IY after statement 1. Either IX or IY can then be taken as the output for integer numbers.

The second and third statements of the subroutine guard against a negative value of IY. The value added at statement 1 is 2^{32}, expressed in this form in order to be able to enter the number. For a 32 bit machine, using the highest order bit as a sign bit, the result of statement 1 is to invert the sign. The two statements following 1 produce the floating-point output. If this is not wanted, the two statements should be omitted to obtain faster execution. The constant in the statement following 2 converts a fixed point number to floating-point.

6-8

Generating Discrete Distributions

We demonstrate first how random numbers are used to generate sequences of numbers from a discrete distribution. When the discrete distribution is uniform, the requirement is to pick one of N alternatives with equal probability given to each. Given a random number r $(0 \leqslant r < 1)$, the process of multiplying by N and taking the integral portion of the product, which is denoted mathematically by the expression $[rN]$, gives N different outputs. The outputs are the numbers $0, 1, 2, \ldots, (N - 1)$. The result can be changed to the range of values C to $N + C - 1$ by adding C.

Alternatively, the next highest integer of the product rN can be taken. In that case, the outputs are $1, 2, 3, \ldots, N$. The next lowest integer can also be taken as output, but note that the rounded value of the product $r \cdot N$ is *not* satisfactory. It produces $N + 1$ numbers as output, since it includes 0 and N, and these two numbers have only half the probability of occurring as the intermediate numbers.

Generally, the requirement is for a discrete distribution that is not uniform, so that a different probability is associated with each output. Suppose, for example, it is necessary to generate a random variable representing the number of items bought by a shopper at a supermarket, where the probability function is the discrete distribution given previously in Table 6-1. A table is formed to list the number of items x and the cumulative probability as shown in Table 6-4.

Taking the output of a uniform random number generator, r, the value is compared with the values of y. If the value falls in an interval $y_i < r \leqslant y_{i+1}$ $(i = 0, 1, \ldots, 4)$, the corresponding value of x_{i+1} is taken as the desired number

Table 6-4   Generating a Non-Uniform
Discrete Distribution

| | | Cumulative |
| No. of Items | Probability | Probability |
x	$p(x)$	y
0	0	0
1	0.10	0.10
2	0.51	0.61
3	0.19	0.80
4	0.15	0.95
5	0.05	1.00

output. The five values of uniformly distributed random numbers derived from Table 6-3 in Sec. 6-5 would, in this case, lead to the five outputs, 2, 2, 1, 5, 2.

Note that it is not necessary that the values of x be in any particular order. Selecting the order carefully can result in more rapid execution of computer routines written to use this method. If the search for the correct interval begins at the value y_1, the average number of comparisons will be minimized if the x values are put in order of decreasing probability.

6-9

Non-Uniform Continuous Random Number Generators

The general requirement in simulation is for a sequence of random numbers drawn from a distribution that is continuous and non-uniform. The commonly used way of deriving such numbers creates the desired number sequence from a uniformly distributed sequence of random numbers.

If $f(x)$ is the desired probability density function, consider the cumulative distribution function, $F(x) = \int_{-\infty}^{x} f(x)dx$. Then $F(x)$ is nondecreasing and lies between 0 and 1. Given a sequence of random numbers, r_i, that are uniformly distributed over the range 0 to 1, each number of the sequence is considered to be a value of the function $F(x)$, and the corresponding value x_i is determined. The sequence of numbers x_i are randomly distributed and have the probability density function $f(x)$. In other words, the inverse of the cumulative distribution function is evaluated with a sequence of uniformly distributed random numbers. The procedure is illustrated graphically in Fig. 6-4. Given a curve displaying $F(x)$ as a function of x, the value of each successive input from the sequence of uniformly distributed random numbers is marked along the vertical axis and the corresponding value of x is determined.

To demonstrate the validity of this result, consider a small interval Δx on the x axis between x and $x + \Delta x$, and the interval ΔF defined by the corre-

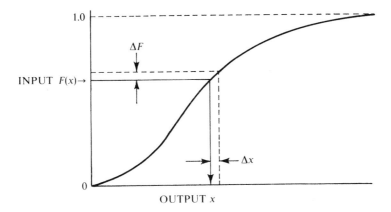

Figure 6-4. Generation of non-uniform continuous random numbers.

sponding points $F(x)$ and $F(x + \Delta x)$, as illustrated in Fig. 6-4. For a number of the output sequence to fall in the interval Δx, the input number must fall in the interval ΔF. Since the input numbers are uniformly distributed over the range 0 to 1, the probability of an input number falling in the interval ΔF is equal to ΔF. The average value of the probability density function for x over the interval Δx is therefore $\Delta F/\Delta x$. In the limit as Δx tends to zero the value of the density function of x is dF/dx which by definition of $F(x)$ is the function $f(x)$.

If the density function $f(x)$ can be described mathematically, it is often possible to find an expression for the inverse of the cumulative distribution function which can then be evaluated with a sequence of uniformly distributed random numbers. Suppose, for example, the desired probability density function is

$$f(x) = a(1 - x) \qquad (0 \leqslant x \leqslant 1)$$
$$= 0 \qquad \qquad \text{elsewhere}$$

Then,

$$F(x) = a\left(x - \frac{x^2}{2}\right)$$

The value of a must be 2 in order to satisfy the condition that $\int_{-\infty}^{\infty} f(x)dx = 1$. Letting $r_i = F(x)$ and solving for x, we find the inverse function:

$$x_i = 1 - (1 - r_i)^{1/2}$$

If r_i is a sequence of uniformly distributed numbers, the numbers x_i are random numbers distributed according to the density function $f(x)$ given above.

Since r_i is uniformly distributed between 0 and 1, so also is $1 - r_i$ and, in this case, it is possible to use the simpler equation

$$x_i = 1 - (r_i)^{1/2}$$

When the probability density function has been derived numerically and the cumulative distribution is approximated by a graph of the type shown in Fig. 6-3, it is necessary to use numerical methods of evaluating the inverse cumulative distribution function. Also, if a mathematically defined density function cannot be inversed, numerical methods must be used.

Suppose the function is represented by a series of straight line segments as is shown in Fig. 6-3. The function is defined at the points $x_i (i = 0, 1, \ldots, N)$ to have the values y_i $(i = 0, 1, \ldots, N)$. The value y_0 is assigned to be 0, and the last defined value, y_N, is assigned to be 1. A table is constructed in which the y_i values are the input of the table, and the values of x_i are the output. This notation is being used to emphasize the fact that it is an inverse function that is being evaluated.

Take the data on telephone call lengths given in Table 6-2 and illustrated as a cumulative distribution in Fig. 6-3. Table 6-5 shows in columns 1 and 2, respectively, the values of y_i and x_i. To evaluate the function between the defined points, an interpolation is made, and for that purpose it is necessary to have the slope of the line, given by the formula

$$a_i = \frac{x_{i+1} - x_i}{y_{i+1} - y_i} \qquad (i = 0, 1, \ldots, N - 1)$$

The values can be calculated when the table is constructed. In this case, they are shown in column 3 of Table 6-5. Note that a_i is the slope of x with respect to y.

To use the table, a sequence of uniformly distributed random numbers, r_j, is matched against the values of y_i. If the input value r_j should equal one of the tabulated values, the corresponding value of x_i is taken. If r_j falls in the interval

$$y_i < r_j < y_{i+1} \qquad (i = 0, 1, \ldots, N - 1)$$

then the output is

$$x_j = x_i + a_i(r_j - y_i)$$

Table 6-5 Generation of Telephone Call Lengths

y_i	x_i	a_i
0	0	10,000.00
0.001	10	10,000.00
0.002	20	10,000.00
0.003	30	5,000.00
0.005	40	1,250.00
0.013	50	357.14
0.041	60	153.85
0.106	70	82.64
0.227	80	57.14
0.402	90	50.76
0.599	100	57.14
0.774	110	82.64
0.895	120	153.85
0.960	130	357.14
0.988	140	1,111.11
0.997	150	5,000.00
0.999	160	10,000.00
1.000	170	—

Taking the 5 random numbers derived from Table 6-3 in Sec. 6-5, it will be found that the corresponding outputs are

r_j	x_j
0.1009	69.23
0.3754	88.48
0.0842	66.65
0.9901	142.33
0.1280	71.82

A number of theoretical distributions frequently occur when describing systems. The next chapter will describe specific methods of generating random numbers from four such distributions. These are the exponential, the normal, the Erlang, and the hyper-exponential distributions. For some other distributions see Ref. (6), which gives rules for computing the generalized exponential (Pearson XI), Weibull, gamma, beta, multivariate normal, and the lognormal distributions.

Exercises

In the following problems that require the use of a uniformly distributed random number generator, use the numbers given in Table 6-3 starting from the top of the leftmost column (column 1) and reading down the column.

6-1 Give the correct value of the constant A that makes the following equations for y a probability density function. Derive formulas for generating random numbers having these distributions and compute the first 10 values.

(a) $y = \dfrac{1}{(x + A)}$ $0 \leqslant x \leqslant 1$

     $= 0$ elsewhere

(b) $y = 0.5 + A(x + 1.5)$ $1 \leqslant x \leqslant 2$

     $= 0$ elsewhere

(c) $y = A \sin x$ $0 \leqslant x \leqslant \pi/2$

     $= 0$ elsewhere

(d) $y = 0.25 + A(x - 1)$ $1 \leqslant x \leqslant 2$

     $= 0.25 - A(x - 3)$ $2 < x \leqslant 3$

     $= 0$ elsewhere

6-2 Approximate the following function with 10 straight lines at equally spaced intervals of x:

$$y = 0.5093 + 0.2 \sin x \qquad 0 \leqslant x \leqslant \pi/2$$

Use the approximation to derive 10 random numbers having this distribution.

6-3 Assuming a normal 365-day year, construct a table from which to generate a number between 1 and 12, inclusive, to represent the month of the year. Assume each day of the year is equally likely to be chosen, and number the months in their normal calendar sequence. Compute the first 10 outputs.

6-4 The probability of a batter striking at a ball is 0.7. When he strikes, the probability of his hitting is 0.6. If he hits the ball, the probability of its being caught is 0.5. Using columns 1, 2, and 3 of Table 6-3 to compute, respectively, the probabilities of striking, hitting, and being caught, determine how many of the first 10 plays will result in a batter being caught.

6-5 A uniformly distributed random integer between 1 and 20, inclusive, is to be generated from each of the numbers in column 1 of Table 6-3. How many of the outputs are prime numbers. (Include 1 as a prime number.)

6-6 Draw a relative frequency distribution from the following data, grouping the data in steps of 200. Find the mean and standard deviation of the data.

854	1128	411	194	2054
174	1268	1105	416	545
565	37	597	224	1180
870	245	559	133	99
416	382	421	3229	943
717	415	662	705	1498
1577	1714	521	484	324
1529	2863	16	151	1967
3073	273	3157	1354	288
1217	93	516	507	723

Bibliography

1 Hammersley, J. M., and D. C. Handscomb, *Monte Carlo Methods*, New York: John Wiley and Sons, Inc., 1964.

2 RAND Corporation, *A Million Random Digits With 100,000 Normal Deviates*, Glencoe, Ill.: The Free Press, 1955.

3 Jansson, Birger, *Random Number Generators*, Stockholm: Almqvist and Wiksell, 1966.

4 Certaine, J., "On Sequences of Pseudo-Random Numbers of Maximal Length," *J. ACM*, V (1958), 353–356.

5 Coveyou, R. R., and R. D. MacPherson, "Fourier Analysis of Uniform Random Number Generators," *J. ACM*, XIV (1967), 100–119.

6 Naylor, Thomas H., Joseph L. Balintfy, Donald S. Burdick, and Kong Chu, *Computer Simulation Techniques*, New York: John Wiley and Sons, Inc., 1966.

7

ARRIVAL PATTERNS
AND SERVICE TIMES

Many systems of interest in a simulation study contain processes in which there is a demand for service that causes congestion. It may, for example, be customers trying to check-out at a supermarket counter, work-pieces waiting for a machine to become available, ships waiting for a berth, and so on. The system can service entities at a rate which, in general, is greater than the rate at which entities arrive, but there are random fluctuations either in the rate of arrival, the rate of service, or both. As a result, there are times when more entities arrive than can be served at one time, and some entities must wait for service. The entities are then said to join a *waiting line* or, more simply, to form a *queue*.

Congestion may be described in terms of three main characteristics. These are

(a) The *arrival pattern*, which describes the statistical properties of the arrivals.
(b) The *service process*, which describes how the entities are served.
(c) The *queuing discipline*, which describes how the next entity to be served is selected.

The service process is described by two main factors: the *service time* and the *capacity*. The service time is the time required to serve an individual entity. The service capacity, or, more simply, the capacity, is the number of entities that can be served simultaneously. A service mechanism that has a capacity of n is also said to have n *channels*.

A third factor that may need to be described in discussing the service is its availability. It may not be available at all times. For example, a machine may break down or be periodically removed from service for inspection. In that case, the availability will be a function of the system conditions. The availability may also be an intrinsic property of the service as occurs, for example, with an elevator which is only able to admit people when it is stopped with its doors open.

To model a system, the probability functions that describe the arrival patterns and the service times must be given. Most system models consist of several activities, interconnected by having the output of one become the input for another. The arrival patterns that result from the transfer of entities between these activities arise from endogenous events and do not have to be described; they are generated as the simulation proceeds. The exogenous arrivals coming to the system from its environment, however, do need to be described.

When measurements have been taken of an arrival pattern, the approximation methods described in Sec. 6-3 can be used to reproduce the distribution. This is likely to be done when studying a specific system. When studying general types of systems, it is desirable to have some fundamental representation of the distribution which, while it may be theoretical, has the general characteristics of the traffic occurring in that type of system. A number of theoretical distributions that serve this purpose will be discussed. In addition, the introduction of stochastic variables into a model makes most of the system performance measures vary stochastically. The measurement of queues is among the most important outputs of a simulation, and the ways of describing and measuring queues will also be discussed.

In simple cases, the statistical properties of the arrivals and the service are independent of time, in which case they are said to be *stationary*. There may, however, be effects that cause these factors to vary with time, in which case the process is said to be nonstationary or *time-variant*. For example, the arrival rate may depend on time of day, resulting in peak load periods, or the rate of service may speed up when there are long queues. Where this happens, the effect is simulated by making some parameter of the distribution, such as the mean, vary according to conditions.

7-2

Arrival Patterns   The usual way of describing an arrival pattern is in terms of the *inter-arrival time*, defined as the interval between successive arrivals. For an arrival pattern that has no variability, the inter-arrival time is, of course, a constant. Where the arrivals vary stochastically, it is necessary to define the probability function of the inter-arrival times. Two or more arrivals may be simultaneous. If n arrivals are simultaneous, then $n - 1$ of them have zero inter-arrival times.

In discussing arrival patterns, the following notation will be used:

T_a Mean inter-arrival time
λ Mean arrival rate

They are related by the equation

$$\lambda = \frac{1}{T_a}$$

Probability distributions have been described so far as either probability density functions or cumulative distribution functions. When describing arrival patterns, it is common practice to express the distribution in terms of the probability that an inter-arrival time is greater than a given time. We define $A_0(t)$ as the *arrival distribution*, so that:

$A_0(t)$ is the probability that an inter-arrival time is greater than t.

Since the cumulative distribution function $F(t)$ is the probability that an inter-arrival time is less than t, it is related to the arrival distribution by

$$A_0(t) = 1 - F(t)$$

From its definition, the function $A_0(t)$ takes a maximum value of 1 at $t = 0$ and it cannot increase as t increases.

7-3
Poisson Arrival Patterns

A common situation is that the arrivals are said to be completely random. Speaking loosely, this means that an arrival can occur at any time, subject only to the restriction that the mean arrival rate be some given value. More formally, it is assumed that the time of the next arrival is independent of the previous arrival, and that the probability of an arrival in an interval Δt is proportional to Δt. If, in fact, λ is the mean number of arrivals per unit time, then the probability of an arrival in Δt is $\lambda \Delta t$. With these assumptions, it is possible to show that the distribution of the inter-arrival times is exponential ((1), p. 13). The probability density function of the inter-arrival time is given by

$$f(t) = \lambda e^{-\lambda t} \qquad (t \geqslant 0) \tag{7-1}$$

It follows that the arrival distribution is

$$A_0(t) = e^{-\lambda t} \qquad (7\text{-}2)$$

The number λ is the mean number of arrivals per unit time. The actual number of arrivals in an interval of time t is a random variable. It can be shown that with an exponential distribution of inter-arrival times, the probability of n arrivals occurring in an interval of length t is given by

$$P(n) = \frac{(\lambda t)^n e^{-\lambda t}}{n!} \qquad (n = 0, 1, 2, \ldots) \qquad (7\text{-}3)$$

This distribution, which is called the *Poisson distribution*, is discrete. The exponential distribution is, of course, continuous, since the inter-arrival time can take any non-negative value. Because of this connection between the two distributions, a random arrival pattern is often called a *Poisson arrival pattern*. Where this term is used, it will mean that the inter-arrival time is exponentially distributed.

The Poisson arrival pattern has great practical importance because it is found to represent arrivals in many types of systems. It also has importance in the theory of queues, because the underlying assumption that an arrival time is independent of the previous arrival enables mathematical solutions to be obtained in a number of cases. The Poisson distribution is not only of importance in discussing arrival patterns, it is one of the most widely used distributions in the application of probability theory and has been used to describe many different phenomena (2). In particular, although the Poisson distribution has been described here as a representation of a random arrival pattern, it is also used to describe a service pattern. When a service time is considered to be completely random, it may be represented by an exponential distribution. In that case, the associated Poisson distribution represents the number of entities served within a given interval.

7-4

The Exponential Distribution

The cumulative distribution function of the exponential distribution is given by

$$y = 1 - e^{-\lambda t}$$

which can be inversed to give

$$\lambda t = -\log(1 - y)$$

where the logarithm is a natural logarithm. Substituting for y, a series of uniformly distributed random numbers between 0 and 1 gives as output a series of random numbers that are exponentially distributed. If the numbers y are uniformly distributed, so also are the numbers $1 - y$, so it is possible to use the simpler formula

$$t = \frac{-\log(y)}{\lambda} = -T_a \log(y)$$

It will be noticed that the exponential distribution is completely characterized by one parameter, its mean value T_a, that appears as a multiplier in the formula for generating exponentially distributed random numbers. Numbers with any mean can be derived from a generator of mean value 1 by multiplying its output by the required mean. This is a specific property of the exponential distribution and cannot generally be applied to other distributions.

Denoting the mean value by AVR, a suitable FORTRAN subroutine using the uniform random number generator described in Sec. 6-7 is

CALL RAND(IX,IY,YFL)

X = −AVR★LOGF(YFL)

The expression LOG(YFL) represents a FORTRAN supplied function for the (natural) logarithm of a floating-point number. The subroutine gives X as a floating-point number.

The appearance of this formula for computing exponentially distributed numbers is deceptively simple. The number of programming statements is small, but the time to execute them can be relatively long. The logarithm function can require a large number of instruction executions, because it is determined from a series that converges slowly. A simulation may need thousands or even hundreds of thousands of random numbers for its completion, so there is a premium on keeping the time required to evaluate a number as small as possible. As a result, many methods have been devised for reducing the time by substituting, either wholly or in part, look-up tables of the type discussed in Sec. 6-9. Naturally, these methods require more storage space in a computer and result in some loss of accuracy; but, where the slight loss of accuracy can be accepted, they can lead to substantial time savings.

As an example, Table 7-1 shows a simple table of values for the function $\log(1 - y)$, (3). Since the table has to be constructed beforehand, there is no advantage in using the alternative form $\log y$ as there would be if the function were being evaluated for each number. The accuracy obtained after multiplying by the required mean value T_a is about 0.1% when $T_a > 250$ and about 1% when $45 < T_a \leqslant 250$.

The approximation in Table 7-1 has arbitrarily truncated the exponential function so that the largest possible value is 8. For more accurate approximations to generate the exponential function, see (4).

Table 7-1   Table for Generating the Exponential Distribution

Input	Output	Slope	Input	Output	Slope
0	0	1.04	0.90	2.30	11.0
0.1	0.104	1.18	0.92	2.52	14.5
0.2	0.222	1.33	0.94	2.81	18.0
0.3	0.355	1.54	0.95	2.99	21.0
0.4	0.509	1.81	0.96	3.20	30.0
0.5	0.690	2.25	0.97	3.50	40.0
0.6	0.915	2.85	0.98	3.90	70.0
0.7	1.20	3.60	0.99	4.60	140
0.75	1.38	4.40	0.995	5.30	300
0.80	1.60	5.75	0.998	6.20	800
0.84	1.83	7.25	0.999	7.0	3,333
0.88	2.12	9.00	0.9997	8.0	—

7-5

The Coefficient of Variation

The question of deciding whether a theoretical distribution can correctly be assumed to represent a particular set of measured data is one that goes beyond the scope of this text. It involves applying a chi-square test, which is discussed in textbooks of engineering statistics. A rough test for data that is non-negative, such as arrival times, can be made by computing the coefficient of variation of the measured data, defined as

$$\text{coefficient of variation} = \frac{\text{standard deviation}}{\text{mean value}} = \frac{\sigma}{T_a}$$

The coefficient of variation provides a measure of the degree to which the data is dispersed about the mean. A coefficient of zero means there is no variation, that is, the data has a constant value. As the coefficient increases, the data becomes more dispersed.

For the case of an exponential distribution of mean value T_a, the standard deviation is also T_a. The coefficient of variation, therefore, is 1. If the coefficient of variation of the measured data is found to be close to 1, it is reasonable to suppose that an exponential distribution may fit the data. The chi-square test should be carried out, however, to test the level of significance to which the data fits. Two other theoretical distributions that are often used when the coefficient of variation is significantly less than or greater than 1 are the Erlang and hyper-exponential distributions, respectively. These will be discussed in the next two sections.

There is a class of distribution functions named after A. K. Erlang, who found these distributions to be representative of certain types of telephone traffic. An arrival pattern governed by this type of function has the following density function and arrival distribution ((1), p. 41):

$$f(t) = (k\lambda)^k \left[\frac{e^{-k\lambda t}}{(k-1)!} \right] t^{k-1}$$

$$A_0(t) = e^{-k\lambda t} \sum_{n=0}^{k-1} \frac{(k\lambda t)^n}{n!}$$

$$(7\text{-}4)$$

where k is a positive integer greater than zero. Figure 7-1 illustrates the Erlang arrival distribution for several values of k. Putting $k = 1$ in Eq. (7-4) and comparing with Eq. (7-1), it can be seen that the case of $k = 1$ is the exponential distribution. The standard deviation of the Erlang distribution is found to be $T_a/k^{1/2}$, so that the coefficient of variation is $1/k^{1/2}$. The coefficient has a maximum value of 1 for the case of the exponential distribution, and it decreases as k increases. It follows that data represented by an Erlang distribution ($k > 1$) clusters more closely to the mean value than exponentially distributed data, so

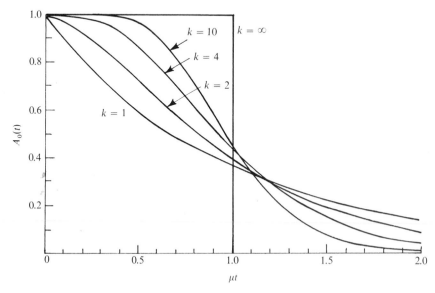

Figure 7-1. Erlang arrival distributions.

that there are fewer low values and high values. Ultimately, as k tends to infinity, the Erlang distribution tends to the case of a constant arrival time, for which the arrival distribution is the step function shown in Fig. 7-1.

When measured data are found to have a coefficient of variation significantly less than 1, it is reasonable to suppose that the data may be represented by the Erlang distribution for which k is closest to the value $(T_a/\sigma)^2$. Again, the significance of the fit should be tested with a chi-square test.

While Eq. (7-4) describing the Erlang density function may appear to be rather complex, the distribution can be given a simple physical interpretation, as illustrated in Fig. 7-2, ((1), p. 44). Suppose there are k stages of service arranged

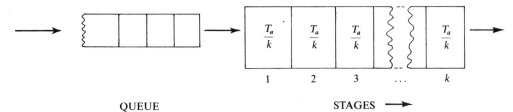

Figure 7-2. Model for Erlang distributions.

in series, each having an exponentially distributed service time with the same mean value T_a/k. When an entity is given service, it passes through all k stages with a random and independently selected service time in each stage. The passage from one stage to the next occurs without loss of time, and the next entity is not allowed to enter the first stage until the preceding entity has cleared all stages. It can be shown that the distribution of over-all service time is an Erlang distribution of the kth order with a mean service time T_a. From this analogy, it is apparent that a number from the Erlang–k distribution with mean value T_a can be created by generating k independent exponentially distributed random numbers, each with mean value T_a/k, and summing the k numbers.

7-7

The Hyper-Exponential Distribution

The hyper-exponential distribution defines a class of distributions that have a standard deviation greater than their mean value; consequently, they can be used to match data found to have a coefficient of variation greater than 1. They represent data where low and high values occur more frequently than with an exponential distribution; the data may in fact be bimodal, showing a peak on either side of the mean value.

As with the Erlang distributions, it is possible to give a physical analogy to explain the nature of the distribution ((1), p. 52). Suppose that there are two parallel stages of processing as shown in Fig. 7-3. Both stages have exponential service times, one with a mean value of $T_a/2s$ and the other with mean value $T_a/2(1 - s), 0 < s \leqslant \frac{1}{2}$. When an entity is to be served, a random choice is made

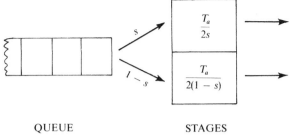

QUEUE   STAGES

Figure 7-3. Model for hyper-exponential distributions.

between the two stages; the probability of choosing the stage with mean of $T_a/2s$ being s and the probability of choosing the other being $(1 - s)$. A second entity cannot enter either stage until the preceding entity has been cleared. The equation for the resultant arrival distribution is found to be ((1), p. 53)

$$A_0(t) = se^{-2s\lambda t} + (1 - s)e^{-2(1-s)\lambda t} \qquad (0 < s \leqslant \tfrac{1}{2})$$

The variance is kT_a^2 so that the coefficient of variation is $k^{1/2}$, where

$$k = \frac{(1 - 2s + 2s^2)}{2s(1 - s)} \qquad (0 < s \leqslant \tfrac{1}{2})$$

When $s = \tfrac{1}{2}$, the value of k is 1 and the distribution becomes the exponential distribution. Figure 7-4 illustrates some hyper-exponential arrival distributions for various values of k. The corresponding values of s are also shown in Fig. 7-4.

Hyper-exponential distributions of random numbers can be generated by the process described in Fig. 7-3. An exponentially distributed random number with mean value of 1 is generated from a uniformly distributed random number. A second uniformly distributed random number is compared to s. If it is less than s, the exponentially distributed random number is multiplied by $T_a/2s$; otherwise, it is multiplied by $T_a/2(1 - s)$.

7-8

Frequently, the service time of a process is constant; but, where it varies stochastically, it must be described by a probability function. In discussing service times, the following notation will be used:

T_s = Mean service time
μ = Mean service rate
$S_0(t)$ = Probability that service time is $> t$
$\rho = \dfrac{T_s}{T_a} = \dfrac{\lambda}{\mu}$

Service Times

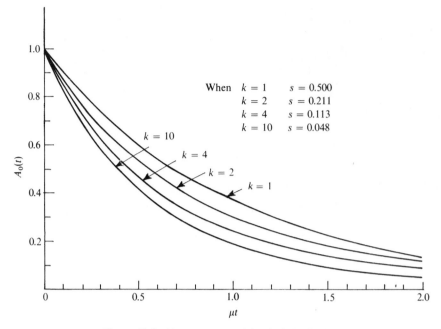

Figure 7-4. Hyper-exponential arrival distributions.

If the service time is considered to be completely random, it may be represented by an exponential distribution or, if the coefficient of variation is found to differ significantly from 1, an Erlang or hyper-exponential distribution may be used. A common situation is that, although the service time should be a constant, there are random fluctuations due to uncontrolled factors. For example, a machine tool may be expected to take a fixed time to turn out a part, but random variations in the amount of material to be removed and the toughness of the material cause fluctuations in the processing time. The normal or Gaussian distribution is often used to represent the service time under these circumstances.

7-9

The Normal Distribution

The normal probability density function is symmetric about its mean value and is completely characterized by its mean value μ and standard deviation σ.

If the transformation $z = (x - \mu)/\sigma$ is made, the distribution is transformed to a form in which the mean is 0 and the standard deviation is 1. In this form, the probability density function is

$$f(z) = \frac{1}{(2\pi)^{1/2}} e^{-z^2/2}$$

It is customary to create generators that determine numbers distributed according to the function $f(z)$ and to derive the variable x that has mean μ and standard deviation σ by the transformation $x = z\sigma + \mu$. Neither the cumulative distribution function nor its inverse can be expressed in terms of simple mathematical functions, but approximation methods are available for their evaluation. A very useful approximation method derives normally distributed random numbers by summing several uniformly distributed random numbers x_i according to the following formula (5):

$$y = \frac{\sum_{i=1}^{k} x_i - \frac{k}{2}}{\left(\frac{k}{12}\right)^{1/2}}$$

The distribution of a sequence of numbers derived from this formula approaches a normal distribution with mean zero and standard deviation 1 as k approaches infinity. Quite small values of k give good accuracy. A convenient value of k is $k = 12$, in which case, the formula takes the form

$$y = \sum_{i=1}^{12} x_i - 6.0$$

This is a direct method of deriving normally distributed random numbers, based upon the properties of random numbers rather than an evaluation of the inverse function. The following FORTRAN subroutine is based on this method and produces a number V which is normally distributed with mean AM and standard deviation S (6):

```
        SUBROUTINE GAUSS(IX,S,AM,V)

        A = 0.0

        DO 50 I = 1,12

        CALL RAND(IX,IY,YFL)

        IX = IY

50      A = A + YFL

        V = (A − 6.0)★S + AM

        RETURN

        END
```

The subroutine is used by issuing the statement

$$\text{CALL GAUSS(IX,S,AM,V)}$$

The subroutine RAND used is the one discussed in Sec. 6-7. If the statement IX = IY is included in RAND, it may be omitted in GAUSS. For other approximation methods of generating normally distributed numbers, see (7) and (8).

A table look-up method can also be used and Table 7-2 is an example of a simple table for generating normally distributed random numbers with mean 0 and standard deviation 1, (3).

The normal distribution assigns a finite probability to all positive and negative values of x. In practice, a quantity being represented by a normal distribution may have only positive values. It may be necessary to test for negative outputs from the generator and arbitrarily set these values to zero. The approximation of Table 7-2 has truncated the normal distribution so that values less than -5 (or greater than $+5$) cannot occur. The computer routine GAUSS cannot produce values of z below -6 (or above $+6$). These values are five and six standard deviations from the mean, respectively. If the mean is less than five standard deviations above zero, Table 7-2 can produce negative values. The routine GAUSS can produce negative values if the mean is less than six standard deviations above zero.

Table 7-2   Table for Generating the Normal Distribution

Input	Output	Slope	Input	Output	Slope
0	−5.0	−33,333	0.57926	0.2	2.63
0.00003	−4.0	−756	0.65542	0.4	2.84
0.00135	−3.0	−206	0.72575	0.6	3.21
0.00621	−2.5	−30.2	0.78814	0.8	3.76
0.02275	−2.0	−11.3	0.84134	1.0	4.59
0.06681	−1.5	−6.22	0.88493	1.2	6.22
0.11507	−1.2	−4.59	0.93319	1.5	11.3
0.15866	−1.0	−3.76	0.97725	2.0	30.2
0.21186	−0.8	−3.21	0.99379	2.5	206
0.27425	−0.6	−2.84	0.99865	3.0	756
0.34458	−0.4	−2.63	0.99997	4.0	33,333
0.42074	−0.2	−2.52	1.0	5.0	—
0.50000	0	2.52			

The third factor for describing congestion is the queuing discipline that determines how the next entity is selected from a queue. The most common queuing disciplines and some of the terms used in describing queues will now be discussed.

(a) A *First-In, First-Out* discipline or, as it is commonly abbreviated, FIFO, occurs when the arriving entities assemble in the time order in which they arrive. Service is offered next to the entity that has waited longest.

(b) A *Last-In, First-Out* discipline, usually abbreviated to LIFO, occurs when entities form a queue in the order in which they arrive, but service is offered first to the one that arrived most recently. This is approximately the discipline followed by passengers getting in and out of a crowded train or elevator. It is the precise discipline for records stored on a magnetic tape that are read back without rewinding the tape.

(c) A *Random* discipline means that a random choice is made between all waiting entities at the time service is to be offered. Unless specified otherwise, the term random implies that all waiting entities have an equal opportunity of being selected.

It has been implied that an entity that joins a queue stays on the queue until it is served. This may not happen; for example, a customer may become impatient and leave. The term *reneging* is used to mean that entities leave the queue, and when this occurs the rules for reneging must be specified. Reneging may depend on the queue length or the amount of time an entity has waited. Frequently, a probability function is given to determine the point of reneging.

When there is more than one queue forming for the same service, the action of sharing service between the queues is called *polling*. Examples of polling systems are a bus stopping along a route to pick up passengers, a clerk circulating around a group of offices to pick up mail, or a computer scanning a number of input terminals to detect the presence of input messages. The polling discipline needs to be specified by giving the order in which the queues are served, the number of entities served at each offering of service, and the time (if any) in transferring service between queues.

Some members of a queue may have priority, meaning that they have a right to be served ahead of all other members of lower priority. There may be several levels of priority and, since there may be several members in the same priority class, a queuing discipline must be specified for the members within a class. A typical situation is for service to be by priority and first-in, first-out within priority class. The priority may be a preassigned attribute of the entity so that it does not depend upon the queue or server, as, for example, the rule women and children first, or it may depend upon the queue. For example, priority may depend upon the length of time an entity has waited in a polling system.

When priority is allowed, a factor that must be considered is what happens when a new arrival has a higher priority than the entity currently being served. The priority may only control the position of the new arrival in the queue. In some cases, however, it may allow the new arrival to displace the entity being served. The new arrival is then said to *interrupt* or *preempt* the service. The term interrupt is usually interpreted as meaning that, when the interrupting entity has finished its service, the service is returned to the entity that was interrupted. Preemption usually means the preempted entity is displaced, either out of the system or back to the queue. However, these terms are not used consistently and their exact meaning should be determined when they are used.

7-11

Measures of Queues

Two principal measures of queues are the mean number of entities waiting and the mean time they spend waiting. Both these quantities may refer to the total number of entities in the system, those waiting and those being served, or they may refer to the entities in the queue only. These quantities are denoted as follows:

L Mean number of entities in the system
L_q Mean number of entities in the queue
W Mean time to complete service (including the wait for service)
W_q Mean time spent waiting in the queue

The probability that an entity will have to wait more than a given time, called the *delay distribution*, is also of interest and will be denoted as follows:

$P(t)$ Probability that the time to complete service is greater than t
$P_q(t)$ Probability that the time spent waiting for service to begin is greater than t.

Figure 7-6 gives examples of delay distributions. From its definition, the function is a maximum at $t = 0$ and it cannot increase with t. The value at $t = 0$, $P(0)$, is of particular interest since it is the probability that some delay, however small, will occur. Correspondingly, $1 - P(0)$ is the probability that there will be no delay. It is, of course, the objective of most studies to minimize the value of $P(0)$. However, a low value of $P(0)$, by itself, may not represent a good system if the "tail" of the distribution stretches out to large values. As a check of this point, the time corresponding to a particular probability level is often picked to indicate how far the tail stretches. For example, from the curve marked $\rho = 0.8$ of Fig. 7-6, it can be seen that, when $\mu = 1$, 30% of the entities can be expected to wait longer than 5 seconds. This type of measure is some-

times called the *grade of service*. Depending upon the desired system performance, the grade of service is specified to be a certain level or certain levels.
For example, a system specification may call for a grade of service in which
not more than 10% of the entities are delayed at all and not more than 0.1%
are delayed more than 2 minutes.

7-12
Mathematical Solutions of Queuing Problems

A number of problems involving queues can be solved mathematically. It is
beyond the scope of this text to discuss these in any detail. For further details,
the reader is referred to (1), (9), or (10). Most of the solutions obtained are for
a single activity servicing exogenous arrivals. Explicit solutions have been
obtained in cases where the arrival pattern and the service times are constant,
exponential, Erlangian, and hyper-exponential. Most activities in a simulation
model receive their input from the output of other activities, and the conditions
of the mathematical solutions are not applicable. Nevertheless, there are times
when the conditions of the mathematical solutions are approximately true, and
the known solutions provide some guidance on the performance to be expected.

By way of illustration, the solution for the case of a single server system with
Poisson arrivals and Erlangian service time will be given, including the special
cases where the Erlang distribution becomes an exponential distribution and
a constant. (See Secs. 7-2, 7-8, and 7-11 for notation.)

The equations for the mean number of entities in the system are ((1), p. 75)

$$L = \frac{2k\rho - \rho^2(k-1)}{2k(1-\rho)} \tag{7-5}$$

$$L_q = L - \rho \tag{7-6}$$

The expected time spent waiting in the system is given by

$$W = \frac{L}{\lambda} = LT_a \tag{7-7}$$

$$W_q = \frac{L_q}{\lambda} = L_q T_a \tag{7-8}$$

For the particular case of an exponential service time, the probability of waiting
in the system is given by ((1), p. 71)

$$P(t) = e^{-(1-\rho)\mu t}$$

$$P_q(t) = \rho P(t)$$

Figure 7-5 shows L, the mean number of entities in the system, and Fig. 7-6 shows $P_q(t)$, the probability of waiting in the queue longer than t, both for the case of an exponential service time.

For the case of hyper-exponential service time, the mean number of entities in the system is given by ((1), p. 84)

$$L = \frac{\rho^2 + \rho(1 - \rho)4s(1 - s)}{4s(1 - s)(1 - \rho)}$$

$$L_q = L - \rho$$

Exercises

7-1 Use Table 7-1 and the first column of Table 6-3 to compute the arrival times of the first 10 arrivals of a Poisson arrival pattern with a mean inter-arrival time of 100. Plot the results as an arrival distribution.

7-2 Calculate the probability of there being n arrivals ($n = 0, 1, \ldots, 10$) in an interval of 10 seconds when the arrivals have a Poisson distribution with a mean value of 0.4.

7-3 Use Table 7-1 and the first four columns of Table 6-3 to generate 20 numbers from a 4th order Erlang distribution with mean value 12. Plot the numbers as an arrival distribution.

7-4 Using the method described in Sec. 7-7, generate 20 numbers with a hyper-exponential distribution having $k = 4$ and a mean value of 10. Use Table 7-1 and column 1 of Table 6-3 to compute the exponentially distributed numbers, and column 2 of Table 6-3 to determine the choice of stages. Plot the results as an arrival distribution.

7-5 Use Table 7-2 and the first column of Table 6-3 to generate 20 numbers having a normal distribution with mean value 50 and standard deviation 5. Plot the distribution of the numbers.

7-6 Determine the coefficient of variation for the following data and judge what distribution best fits the data. Plot the data as an arrival distribution :

95	52	79	31	79	140	175	101	140	158
80	92	90	67	88	97	87	98	105	80
103	106	114	115	100	132	112	162	104	124
116	81	197	138	100	44	156	92	72	97
88	69	117	64	51	138	60	67	87	78

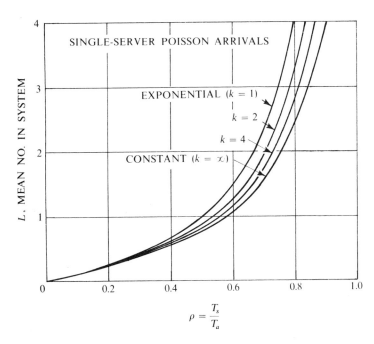

Figure 7-5. Number of entities in a system with Erlang service distribution.

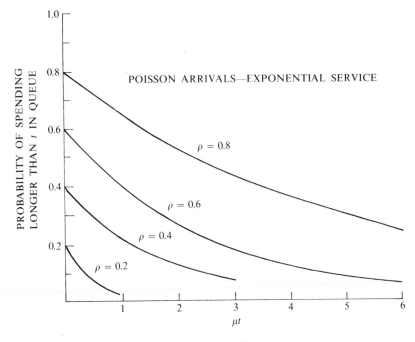

Figure 7-6. Probability of spending time on a queue.

Bibliography

1 Morse, Philip M., *Queues, Inventories and Maintenance*, New York : John Wiley and Sons, Inc., 1958.

2 Haight, Frank A., *Handbook of the Poisson Distribution*, New York : John Wiley and Sons, Inc., 1967.

3 General Purpose Systems Simulator II, Form No. H20-6346, IBM Corp., White Plains, N. Y.

4 MacLaren, M. D., G. Marsaglia, and T. A. Bray, "A Fast Procedure for Generating Exponential Random Variables," *Comm. ACM*, VII, No. 5 (1964), 298–300.

5 Hamming, R. W., *Numerical Methods for Scientists and Engineers*, New York : McGraw-Hill Book Company, 1962, p. 34.

6 System/360—Scientific Subroutine Package, Form No. H20-0205, IBM Corp., Data Processing Division, White Plains, N. Y.

7 Burr, Irving W., "A Useful Approximation to the Normal Distribution With Application to Simulation," *Technometrics*, IX, No. 4 (Nov. 1967), 647–651.

8 Marsaglia, G., M. D. MacLaren, and T. A. Bray, "A Fast Procedure for Generating Normal Random Variates," *Comm. ACM*, VII, No. 1 (1964), 4–9.

9 Saaty, T. L., "Résumé of Useful Formulas in Queuing Theory," *Operations Research*, V (Apr. 1957), 162–187.

10 Cox, D. R., and Walter L. Smith, *Queues*, New York : John Wiley and Sons, Inc., 1961.

8

DISCRETE SYSTEM SIMULATION

Discrete systems were characterized in Chap. 1 as systems in which the changes are predominantly discontinuous, and the term *event* was used to describe the occurrence of a change at a point in time. An event may cause a change in the value of some attribute of an entity, it may create or destroy an entity, or it may start or stop an activity. Since the technique of simulation consists of following the changes in a model of a system, the task of simulating discrete systems requires the construction of a program in which it is possible to follow the sequence of events.

Two general points of view may be taken of the way in which discrete events are identified (1). In one point of view, referred to as *particle-oriented*, or *material-based*, attention is focused on the entities in the system, and the simulation is regarded as the task of following the changes that occur as the entities move from activity to activity. In that case, the times at which system changes occur are treated as attributes of the entities. In the other point of view, referred to as *event-oriented*, or *machine-based*, attention is focused on the activities, and the simulation follows the history of the activities as they are applied to different entities. In that case, the times at which system changes occur are treated as characteristics of the activities.

Whichever point of view is taken, the execution of a simulation remains the same. Records must be kept of all the activities in progress and the entities involved, and they must be periodically changed to reflect the sequence of events within the system. To do so, records must be kept of *event times* and calculations must compute future event times as the simulation proceeds.

8-2

Representation of Time

The passage of time is recorded by a number referred to as *clock time*. It is usually set to zero at the beginning of a simulation and subsequently indicates how many units of simulated time have passed since the beginning of the simulation. Unless specifically stated otherwise, the term *simulation time* means the indicated clock time and not the time that a computer has taken to carry out the simulation. As a rule, there is no direct connection between simulated time and the time taken to carry out the computations. The controlling factor in determining the computation time is the number of events that occur. Depending upon the nature of the system being simulated, and the detail to which it is modeled, the ratio of the simulated time to the real time taken can vary enormously. If a simulation were studying the detailed workings of a digital computer system, where real events are occurring in time intervals measured in fractions of microseconds, the simulation, even when carried out by a high speed digital computer, could easily take several thousand times as long as the actual system operation. On the other hand, for the simulation of an economic system, where events have been aggregated to occur once a year, a hundred years of operation could easily be performed in a few minutes of calculations.

Two basic methods exist for updating clock time. One method is to advance the clock to the time at which the next event is due to occur. The other method is to advance the clock by small (usually uniform) intervals of time and determine at each interval whether an event is due to occur at that time. The first method is referred to as *event-oriented*, and the second method is said to be *interval-oriented*. Discrete system simulation is usually carried out by using the event-oriented method, while continuous system simulation normally uses the interval-oriented method.

It should be pointed out, however, that no firm rule can be made about the way time is represented in simulations for discrete and continuous systems. An interval-oriented program will detect discrete changes and can therefore simulate discrete systems, and an event-oriented program can be made to follow continuous changes by artificially introducing events that occur at regular time intervals.

8-3

Generation of Arrival Patterns

An important aspect of discrete system simulation is the generation of exogenous arrivals. It is possible that an exact sequence of arrivals has been specified for the simulation. The sequence may, for example, be the results of some observations on the system. Further, when there is no interaction between the exogenous arrivals and the endogenous events of the system, it is permissible to create a sequence of arrivals in preparation for the simulation. Usually, however, the simulation proceeds by creating new arrivals as they are needed.

The exogenous arrival of an entity is defined as an event and the arrival time of the next entity is recorded as one of the event times. When the clock time reaches this event time, the event of entering the entity into the system is executed and the arrival time of the following entity is immediately calculated from the inter-arrival time distribution. The term *boot-strapping* is often used to describe this process of making one entity create its successor. The method requires keeping only the arrival time of the next entity; it is, therefore, the preferred method of generating arrivals for computer simulation programs.

The arriving entity usually needs to have some attribute values generated, in which case, attention must be paid to the time at which the values are generated. They could be generated at the time the arrival time is calculated, or they could be generated when the entity actually arrives. If there is no inter-action between the attributes and the events occurring within the system, the generation may be done at either time. If, however, the attribute values depend upon the state of the system, it must be remembered that, at the time of generating the arrival time, the actual arrival is still an event in the future. The generation of the attribute values must then be deferred until the arrival event is executed. For example, a simulation is discussed in the next section in which telephone calls are generated. The call length and the origin of the call need to be generated. There is no interaction between the distribution of call length and the state of the system, so the call length can be generated at the time the arrival time is decided or when the call arrives. However, a call cannot come from a line that is already busy, so the choice of origin must be left until the call arrives. To select the origin when the arrival time is decided carries the risk that some other call will have made the proposed origin busy before the call arrives.

8-4
Simulation of a Telephone System

To illustrate the principles involved in the simulation of a discrete system, consider the example of a simple telephone system illustrated in Fig. 8-1, (2). The system has a number of telephones (only the first eight are shown), con-nected to a switchboard by lines. The switchboard has a number of links which can be used to connect any two lines, subject to the condition that only one connection at a time can be made to each line. It will be assumed that the system is a lost call system, that is, any call that cannot be connected at the time it arrives is immediately abandoned. A call may be lost because the called party is engaged, in which case the call is said to be a busy call; or it may be lost because no link is available, in which case it is said to be a blocked call. The

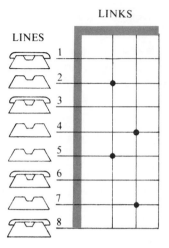

Figure 8-1. Simple telephone system.

object of the simulation will be to process a given number of calls and determine what proportion are successfully completed, blocked, or found to be busy calls.

The current state of the system, shown in Fig. 8-1, is that line 2 is connected to line 5, and line 4 is connected to line 7. One way of representing the state of the system is shown in Fig. 8-2. Each line is treated as an entity, having its availability as an attribute. A table of numbers is established to show the current

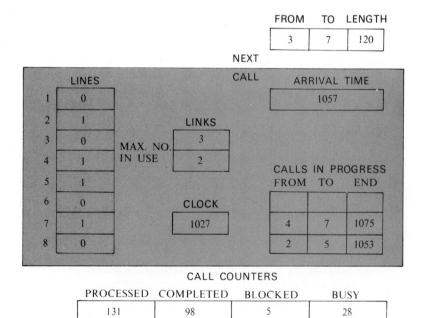

Figure 8-2. System state-1.

status of each line. A *zero* in the table means the line is free, while a *one* means that it is busy. It is not necessary that a detailed history be kept of each individual link, since each is able to service any line. It is only necessary to incorporate in the model the constraint imposed by the fact that there is a fixed number of links (in this case, three). Under these circumstances, the group of links is represented as a single entity, having as attributes the maximum number of links and the number currently in use. Two numbers, therefore, represent the links.

To keep track of events, a number representing clock time is included. Currently, the clock time is shown to be 1027, where the unit of time is taken to be 1 second. The clock will be updated in the event-oriented manner as the simulation proceeds. Each call is a separate entity having as attributes its origin, destination, and length. The simulation will be carried out using the particle-oriented view of events, so that it will be necessary to generate, as a further attribute of the call, the time at which the call finishes. There is a list of *calls-in-progress* showing which lines each call connects and the time the call finishes. To generate the arrival of calls, the bootstrap method described in the previous section is used, so that a record is kept of the time the next call is due to arrive. It will be assumed that the call is equally likely to come from any line that is not busy, and that it can be directed to any line, other than itself, irrespective of whether that line is busy or not. The choice of origin must be left until the call arrives. For convenience, the origin, destination, and call length will all be generated at that time. To help explain the action of the simulation, these choices are shown in Fig. 8-2, although, in practice, they would not have been generated until the clock reaches 1057, the time of the next arrival. The generation of the call length can, in fact, be deferred until not only has the call arrived but it is determined that it can be connected.

The set of numbers within the main box of Fig. 8-2 records the state of the system at time 1027. There are two activities causing events; new calls can arrive and existing calls can finish. As shown in Fig. 8-2, there are three future events; the call between lines 2 and 5 is due to finish at time 1053, the call between lines 4 and 7 is due to finish at time 1075, and a new call is due to arrive at time 1057.

The simulation proceeds by executing a cycle of steps to simulate each event. The first step is to *scan* the events to determine which is the next potential event. In this case, the next potential event is at 1053. The clock is updated, and the second step is to *select* the activity that is to cause the event. (In an event-oriented organization, the event time is a characteristic of an activity and this step would identify the entity to be acted upon.) In this case, the activity is to disconnect a call. There are no conditions to be met when a call is disconnected, so the event will be executed; but, in general, the third step is to *test* whether the potential event can be executed. The fourth step is to *change* the records to reflect the effects of the event. The call is shown to be disconnected by setting to zero the numbers in the lines table for lines 2 and 5, reducing the number of links in use by 1, and removing the finished call from the calls-in-progress table. As a fifth and final step in the execution cycle, it may be necessary to *gather* some statistics for the simulation output. Counters are set aside to record the number of calls that have been processed, completed, or lost through being blocked or busy. With the disconnection of a call, the counts of processed calls and of completed calls are both increased by one. The state of the system then appears as shown in Fig. 8-3. Assuming the simulation is to continue, the cycle of actions just described is repeated.

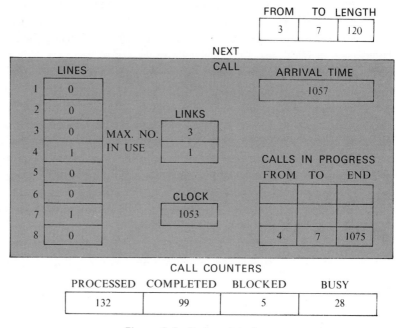

Figure 8-3. System state-2.

It can be seen that the next event is the arrival of a call at time 1057. The clock is updated to the arrival time and the attributes of the new arrival are generated. Since the selected activity is to connect a call, it is necessary to carry out tests; first to see if a link is available, then to see if the party is busy. In this case, the called party is busy so the call is lost. Both the processed calls and the busy calls counters are increased by one. A new arrival is generated and the state of the system at the time the call is lost then appears as shown in Fig. 8-4. Suppose the next arrival time is 1063, and that when this arrival occurs the call will be from line 3 to 6 and will last 98 seconds. Again, the next potential event is an arrival, but this time the arriving call can be connected so the state of the system moves to that shown in Fig. 8-5.

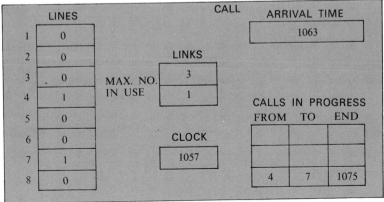

Figure 8-4. System state-3.

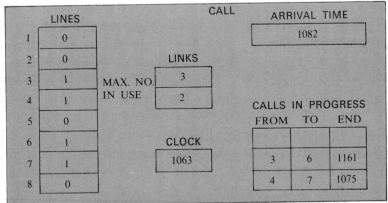

Figure 8-5. System state-4.

The procedure will be repeated until some limit on the length of the simulation is reached. Typically, the simulation will be run until a given number of calls has been processed or until a certain time has elapsed.

8-5

Simulation Programming Tasks

Having demonstrated with a particular example the way a discrete simulation proceeds, it is possible to outline in general the tasks involved in preparing a computer program for a simulation. There are three main tasks to be performed, as shown in Fig. 8-6. The first task is to generate a model and initialize it. From the description of the system, a set of numbers must be created to represent the state of the system. This set of numbers will be called the *system image* since its purpose is to reflect the state of the system at all times. The activities of the system must be represented as routines that are to carry out the changes to the system image. The second task is to program the procedure that executes the cycle of actions involved in carrying out the simulation. This procedure is referred to as the *simulation algorithm*. While the routines representing the system activities are specific to the system being simulated, the simulation algorithm need not be. The third task is the generation of an output report. The statistics gathered during the simulation will usually be specified by the report generator.

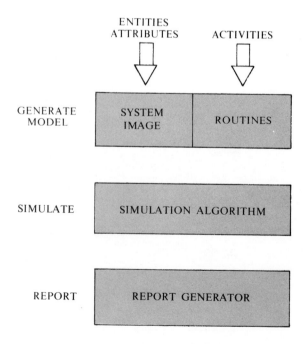

Figure 8-6. Simulation programming tasks.

The general flow of control during the execution of a simulation program is illustrated in Fig. 8-7. Shown at the top of the figure is the task of generating the model, which is executed once. At the bottom is the report generation task, which is usually executed once at the end of the simulation, although it is not unusual to print out intermediate results as the simulation proceeds. Carrying

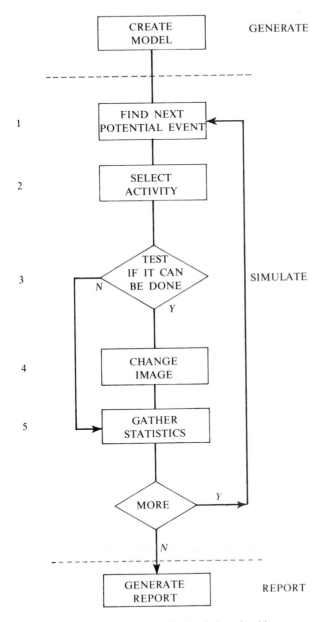

Figure 8-7. Execution of a simulation algorithm.

out the simulation algorithm involves repeated execution of the five steps described previously. These steps are:

1. *Find* the next potential event.
2. *Select* an activity.
3. *Test* if the event can be executed.
4. *Change* the system image.
5. *Gather* statistics.

In the example of the telephone system, a call is abandoned if it cannot be immediately connected. In general, however, when a test shows that an event cannot be executed, the event will wait for conditions to become favorable for its execution. The simulation algorithm must then arrange for the delayed event to appear at some future time when conditions have changed.

8-6
A Simulation Program Flow Chart

To carry the telephone system example further we will produce a flow chart showing how a computer simulation program could be organized. As before, it will be assumed that calls that cannot be connected will be lost; and the purpose of the simulation will be to process N calls, count how many calls are completed, how many are lost because the links are busy, and how many are lost because the called party is busy. Several runs are planned, varying the maximum number of links in the system. A flow chart for the program is shown in Fig. 8-8.

When the program is started, the model will be initialized for the particular run to be made. Included in the input information will be the number of links and the total number of calls to be simulated. The program will read the input data from cards and place them in their appropriate places. An initialization routine sets various numbers to their correct initial values; for example, it will zero the clock time and the counters. The first arrival time is generated and the program enters the main routine concerned with executing each cycle of events. This initial entry point is also the point from which to recycle when all changes have been executed for one event.

Suppose the next call finishes at T_F and the next arrival is at T_A. The program first decides whether the next activity is to disconnect a call or attempt to connect a call. The test of the condition $T_F \leq T_A$ makes this choice. Note that with the test defined this way, a coincidence of a disconnect and a new arrival will result in the disconnect being serviced first. Whichever activity is decided upon, the program updates the clock. When two events are coincidental, the program must execute them successively and the clock will remain unchanged until all events at that time have been executed.

In the case of disconnecting a call, the activity is unconditional and the program immediately carries out the necessary action. This will be to make the

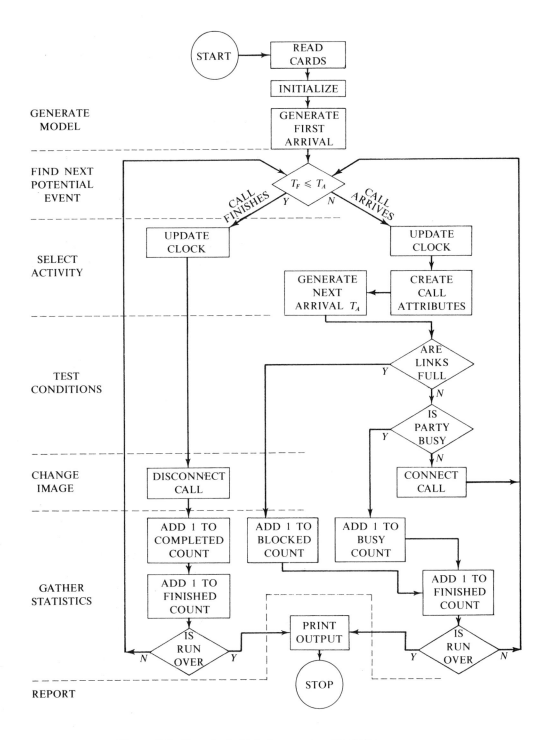

Figure 8-8. Flow chart of telephone system simulation.

type of changes made between Figs. 8-2 and 8-3. Since a call has just finished, the time the next call finishes, T_F, is updated. The program collects statistics and returns for another cycle. In the case of a new call arriving, the attributes of the call are created at that time and the next arrival time is generated. Tests are then made to see if the call will be connected or lost either because no link is available or the called line is busy. When a call is connected, the program must adjust the records by entering the new call in the list of calls in progress. This corresponds to the changes made between Figs. 8-4 and 8-5. When a call is lost, the appropriate statistics must be gathered.

The program checks to see whether enough calls have been completed to finish the simulation run. If not, the program resumes for one more cycle. When enough calls have been completed, the simulation is stopped, the results are printed, and the program terminates. As indicated in Fig. 8-8, the program can be divided into broad sections of scanning events, selecting an activity, testing, changing the system image, and gathering statistics.

8-7
Discrete Simulation Languages

A number of programming languages have been produced to simplify the task of writing discrete system simulation programs. A list of 22 such languages will be found in (3), which also contains an extensive review of six of the languages. Other reviews of simulation languages will be found in (4), (5), and (6).

Essentially, these programs embody a language with which to describe the system, and a programming system that will establish a system image and execute a simulation algorithm. Each language is based upon a set of concepts used for describing the system. The term *world-view* has come to be used to describe this aspect of simulation programs (5). The user of the program must learn the world-view of the particular language he is using and be able to describe the system in those terms. Given such a description, the simulation programming system is able to establish a data structure that forms the system image. It will also compile, and sometimes supply, routines to represent the activities. Routines are supplied to carry out such functions as scanning events, updating the clock, gathering statistics, and maintaining events in time and priority sequence. These are needed to effect the simulation algorithm. Most programs also provide a report generator.

There is, however, a great variety in both the world-views of the languages and the degree to which the programming systems relieve the user of programming details. In general, most languages view the world in terms of entities with attributes and activities, as these concepts were introduced in Chap. 1. Several languages put particular emphasis on sets of entities to the point where a set is regarded as a fundamental element of a system description. Some introduce the concept of a *process* as a conglomeration of activities that can be used to describe a major sector of the model (7).

It is not feasible to discuss here all the simulation languages. Instead, the discussion will be limited to two languages, GPSS and SIMSCRIPT. Following the next two chapters, which discuss the use of FORTRAN for simulation, there are two chapters devoted to GPSS and a further two chapters discussing SIMSCRIPT. The reasons for choosing these two languages are that GPSS and SIMSCRIPT are the two most widely used discrete simulation languages, and they illustrate the wide divergence that exists among simulation languages. GPSS is a particle-oriented program while SIMSCRIPT is event-oriented. For further discussion of the principles of discrete simulation languages, see Ref. (8).

GPSS has been written specifically for users with little or no programming experience, while SIMSCRIPT, along with most other simulation languages, requires programming skill to the level where the user is able to program in FORTRAN or ALGOL. The simplifications of GPSS result in some loss of flexibility; so that, while SIMSCRIPT requires more programming skill, it is capable of representing more complex data structures and can execute more complex decision rules. This difference might be expected to affect the range of application of the languages; however, the survey of Ref. (3) finds no evidence that either language is restricted to any particular types of system. Both GPSS and SIMSCRIPT appear to be general enough to be equally applicable to a wide variety of systems. The differences are more accurately summed up by saying that the greater programming flexibility of SIMSCRIPT means that, in more complex models, SIMSCRIPT is able to produce a more compact model that requires less storage space and, generally, will be executed more rapidly. A comparison of these two languages will now be given indicating broadly how they perform the main programming tasks illustrated in Figs. 8-6 and 8-7. The reader may find it helpful to return to this section after reading the chapters concerned with GPSS and SIMSCRIPT.

GPSS is a block diagram language. A number of basic block types are defined, each representing a particular action. The user must describe the activities of the system in terms of these basic block types. As a matter of programming convenience, most simulation languages distinguish between temporary and permanent entities. Temporary entities are created during the running of the simulation, move through the system, and are ultimately removed from the system. Permanent entities are system elements, such as items of equipment, that remain throughout the simulation. In GPSS, temporary entities are identified as transactions which can carry parameters representing attributes. Permanent entities are identified with pre-defined concepts such as facilities, storages, and logic switches. The world-view of GPSS, therefore, requires that all the entities of the system be identified with predetermined entities of the program. The activities must be described in terms of the fixed block types. The sequence of activities is established by the block diagram indicating how the blocks are interconnected and, where necessary, the conditions for a choice between alternative paths.

With this formal world-view, GPSS is able to establish a system image as a set of well-defined tables, as indicated in Fig. 8-9. Records representing the transactions are created according to the specifications of one of the block types. Transactions move from block to block at certain times controlled by the blocks. Where the transactions are unable to move at their allotted time because of existing conditions, GPSS is able to identify what condition of the program entities is causing the delay, and can automatically execute the delayed event when the condition is seen to change. The selection of the next activity is carried out by interpreting the block diagram to find out what block the transaction is due to enter. Since all system entities have been identified in terms of standard program entities, testing whether an event can be executed becomes an interpretation of the current status of the entity tables. Similarly, since all activities have been translated into standard block types, the execution of an activity only requires transfer of control to a pre-written block type routine.

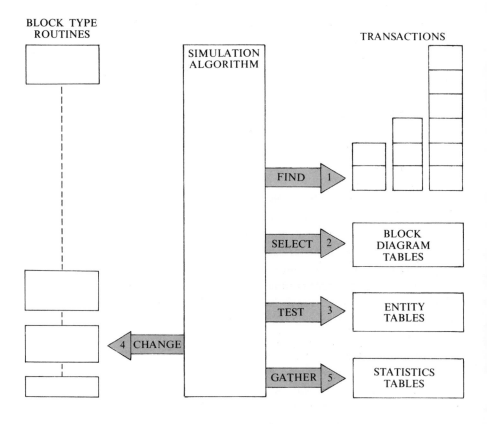

Finally, the gathering of statistics is controlled by certain of the block types together with some internal routines. A fixed report generator gives statistics on the movement of transactions and the use made of equipment, although some editing of the output with explanatory comments is possible.

The SIMSCRIPT world-view also represents the system in terms of entities, both temporary and permanent, and activities. Each type of entity, however, must be individually defined by giving names to the entities and to their attributes, and specifying their data structure. The activities are represented by individually written routines (called event routines), written in SIMSCRIPT language. The linking of activities is carried out by event notices which are also individually defined by the user. Each event notice corresponds to an event due to occur at a particular time. It identifies the activity (event routine) to be activated and is generally destroyed when the activity commences. Within the event routine activated by an event notice, the user must create further event notices for scheduling follow-on events.

The execution of the simulation algorithm in SIMSCRIPT is illustrated in Fig. 8-10. Event notices are automatically kept in time order, and the process of scanning for the next event is one of maintaining this file and selecting the first notice. Since an event notice names the event routine it is to activate, the

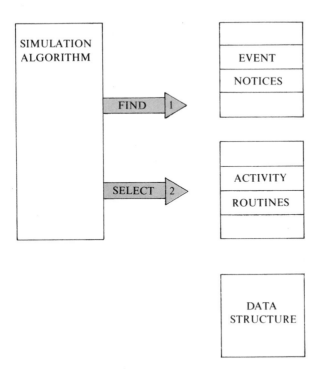

Figure 8-10. Execution of simulation in SIMSCRIPT.

second step, selecting the correct activity, transfers to the indicated event routine. The remaining three steps of testing whether the event can be executed, changing the system image, and gathering statistics are programmed within each individual event routine. This is illustrated in Fig. 8-11, which shows the structure of a typical routine. A number of simulation-oriented statements within SIMSCRIPT are available. In particular, statements for entering and removing temporary entities from sets are available, and the gathering of statistics is programmed directly. Where an event cannot be executed at its scheduled time, it is the responsibility of the user to file away the delayed event and reactivate it when conditions are found to be favorable.

A separate form is used to describe the output required from the program, allowing extensive editing of the results. Tables and counters used to collect the statistics are defined by the user as attributes of the system entities associated with the statistics.

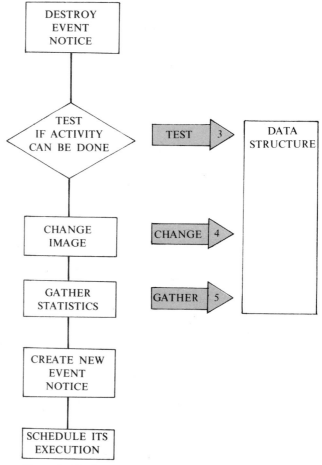

Figure 8-11. Execution of an event routing in SIMSCRIPT.

8-1   Recalculate the telephone system simulation, assuming the inter-arrival time is exponentially distributed and the call lengths are normally distributed. The mean inter-arrival time is 20 and the calls have a mean length of 80 with a standard deviation of 20. Use Table 7-1 and column 1 of Table 6-3 to generate the call arrivals, and use Table 7-2 and column 2 of Table 6-3 to generate the call lengths. Assume the system starts with a call from line 2 to line 5, due to end at time 70, and a call from 4 to 7 to end at 150. Assume that there are 8 lines and 3 links. The origins and destinations of the first five calls are

Call Number	Origin	Destination
1	1	6
2	3	8
3	1	8
4	5	1
5	2	5

How many of the first five calls are blocked or find busy lines?

8-2   Consider the market model of Sec. 2-5 for the stable conditions (case a). Suppose the market is not cleared; instead, the supplier only sells $N\%$ of his goods, where N is a uniformly distributed random number between 80 and 99, inclusive. The price he receives is determined by the point of the demand curve that corresponds to the amount sold. The supplier continues to determine the supply from the last market price. Use the random numbers in column 1 of Table 6-3 to simulate the first 6 periods of the market.

8-3   The birth rate of a population is normally distributed with mean value 5% and standard deviation 0.5%. The death rate is also normally distributed with mean 6% and standard deviation 1%. Use Table 7-2 and columns 1 and 2 of Table 6-3 to compute births and deaths, respectively, and calculate the population changes for 5 years at 1 year intervals.

8-4   A reservoir holds 1,000,000 gallons of water, and it is drained at a steady rate of 10,000 gallons a day. Rainfall occurs with a Poisson distribution with a mean rate of 1 in 10 days. The amount of rain captured by the reservoir is normally distributed with a mean of 80,000 gallons and a standard deviation of 5,000 gallons. Use Tables 7-1 and 7-2 and columns 3 and 4, respectively, of Table 6-3, to compute the time and quantity of rainfalls. Find the contents of the reservoir after 60 days. Assume the reservoir starts full and that water exceeding the capacity of the reservoir is lost.

8-5 Cars arrive randomly at a toll booth and pay tolls. If necessary, they wait in a queue to be served in order of arrival. The inter-arrival time, measured to the nearest second, is found to be uniformly distributed between 0 and 9, inclusive. The time to pay is also random and between 0 and 9 seconds, but with the following distribution:

$$f(t) = \frac{t^{-1/2}}{6} \qquad 0 \leqslant t \leqslant 9$$

$$= 0 \qquad \text{elsewhere}$$

Generate the arrival times from column 5 and the paying times from column 6 of Table 6-3. Truncate all derived numbers to the nearest integer. Find the arrival times and the times taken to pay the toll for the first five cars. At what time does the fifth car clear the toll booth?

Bibliography

1 Tocher, K. D., "Some Techniques of Model Building," *Proceedings of the IBM Scientific Computing Symposium: Simulation Models and Gaming*, IBM Corp., Data Processing Division, White Plains, N. Y., 1966, 117–155.

2 Blake, K., and G. Gordon, "Systems Simulation With Digital Computers," *IBM Systems Journal*, III, No. 1 (1964), 14–20.

3 Teichroew, Daniel, and John Francis Lubin, "Computer Simulation: Discussion of the Technique and Comparison of Languages," *Comm. ACM*, IX, No. 10 (Oct., 1966), 723–741.

4 Tocher, K. D., "Review of Simulation Languages," *Operations Research Quarterly*, XVI, No. 2 (June, 1964), 189–217.

5 Krasnow, H. S., and R. Merikallio, "The Past, Present and Future of General Simulation Languages," *Management Science*, XI (Nov., 1964), 236–267.

6 Freeman, D. E., "Programming Languages Ease Digital Simulation," *Control Engineering* (1964), 103–106a.

7 Parente, R. J., and H. S. Krasnow, "A Language for Modeling and Simulating Dynamic Systems," *Comm. ACM*, X, No. 9 (Sept., 1967), 559–567.

8 Buxton, J. N. (ed.), *Simulation Programming Languages*, Amsterdam: North-Holland Publishing Co., 1968.

9

SIMULATION WITH FORTRAN

Although many simulation languages are available, many simulation studies are carried out without using them. In particular, FORTRAN is widely used, since it is a well-known and readily available general-purpose language for programming scientific and engineering problems. Probably the most common reason for not using a simulation language is that a programming system using the language is not available to the person conducting the simulation study. It may also be that he has not the time or inclination to learn the language. Sometimes, it is felt that a model is too large or too complicated to be programmed with a simulation language, or that, if it can be programmed, the program will be inefficient.

A number of attempts have been made at comparing simulation programs written in various languages, including FORTRAN; see, for example, Refs. (1), (2), and (3). It proves to be difficult to draw comparisons, since the results depend upon the nature of the model and the skill of the programmer. There seems to be little doubt that general-purpose simulation languages, such as CSMP, GPSS, and SIMSCRIPT, collectively, have the capability for programming virtually all types of model. At the same time, it is apparent that the general-purpose nature of these simulation languages, or, more precisely, the programming systems used to implement them, cause some loss of efficiency in execution time and in the use of storage space. In particular, models requiring that very large quantities of data be available at any time can present difficulties when using simulation languages.

The matter of execution speed can easily be viewed out of perspective. Section 2-3 discussed the steps involved in carrying out a simulation study, and it was seen that the actual running of a simulation program is but one step in a chain of many. Usually, the most significant factor controlling the length of a simulation study is the time needed to construct a model and produce a working program. The help that can be given here by a simulation language can greatly outweigh any time saving achieved by programming in a non-simulation language. Where the model is to be run many times and will remain essentially unchanged throughout a study, the saving of time that can be gained may justify the effort of writing programs in a non-simulation language.

For these reasons, and also for the purpose of demonstrating the programming techniques used to implement simulation languages, this chapter and the following will demonstrate the use of FORTRAN in simulation.

9-2
A FORTRAN Simulation of the Telephone System

The system to be studied is the same telephone system studied in the last chapter. A number of subscribers are connected by lines to a switchboard, which can use links to connect any two subscribers. The program will be organized so that the number of subscribers and links can be set to different values, although an arbitrary maximum of 100 lines will be made. In addition, the calling rate and the total number of calls to be run will be variable. An input card will supply the information for each run, and the program will be organized so that several runs can be made with one loading of the program, by supplying one card for each run. There will be other types of input cards, so each card carries a card type number. One card type, number 2, will be explained in the next chapter. All fields are punched with positive integral numbers and the numbers must be right-justified. A RUN card, type 1, has the format shown in Table 9-1:

Table 9-1   Run Card Format

Field No.	Columns	Information
1	1–6	Card type = 1
2	7–12	No. of lines
3	13–18	No. of links
4	19–24	Mean inter-arrival time
5	25–30	Mean call length
6	31–36	No. of calls to be run
7	37–42	Control = 0, print
		= 1, no print

If the control character in field 7 is a 0 (or is left blank), it indicates that results are to be printed at the end of the run. A 1 (or any non-zero integer) indicates no printing. This usually means the run is for the purpose of initially loading the system in preparation for a run from which statistics will be gathered. The program will run the simulation but will not print results.

A RESET card, which has the number 3 in field 1, instructs the program to start a run from the point at which the system was left at the end of a previous run. It must follow a RUN card, with a print control character of either 0 or 1. The action taken by the program on reading a RESET card is to wipe out all statistics gathered in the preceding run, reset the clock to zero but leave all existing calls exactly as they were at the end of that run. The program then restarts, runs for the number of calls indicated in field 6 of the RESET card, and prints results. An END card, with the number 4 in field 1 and nothing else, is placed at the end of the input deck, even when there is only one run. Instructions for the input cards form the first part of the listing in Fig. 9-11.

It will be assumed that the system is a lost call system; that is, a call that cannot be connected will be abandoned. The call may be lost because the called line is engaged, in which case it is a busy call; or it may be lost because all links are in use, in which case it is a blocked call. If both busy and blocked conditions exist simultaneously, the call will be treated as a blocked call, since this is the first condition that is tested. An output is printed after each run to give the following information:

(a) Clock time at which the simulation stopped
(b) Number of calls successfully completed
(c) Number of busy calls
(d) Number of blocked calls

To avoid mixing up outputs from different runs, the output report will always begin by repeating the information on the input cards.

9-3
Organization of the Program

The flow chart of the complete program is the same as the flow chart given in Fig. 8-8, developed for carrying out this same simulation by hand.

When a RUN card has been read, the program is initialized, the first arrival time is generated, and the control passes to a major loop that processes each event. A check is first made to see whether the next event is the arrival of a new call or the termination of an existing call. Whichever event occurs, the clock is updated accordingly. Having selected the next event, the program tests whether the event can be executed. For a terminating call, there are no conditions to be met, so the program proceeds directly to a routine that disconnects the call. When a new call arrives, the program creates the attributes of the call and generates the time of the next arrival. Tests are then made to see if the call can be connected; if so, the connection is made. If not, the program collects some statistics and checks whether more calls should be run. The program then returns to find the next event; otherwise, it computes some final statistics and prints a report.

9-4
Construction of the System Image

The system image is a set of tables and numbers that record the state of the system. The one that will be constructed follows the general pattern of the one described in Sec. 8-4 and illustrated in Fig. 8-2. It consists of the tables and locations shown in Fig. 9-1.

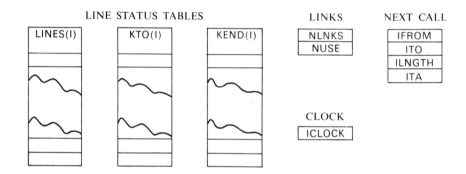

Figure 9-1. System image.

Although the number of calls in progress varies, for simplicity they are recorded in three fixed size tables having one word for each line. This organization wastes space and an alternative way of organizing records whose number fluctuates will be discussed in the next chapter. A table called LINES is used to record whether a line is free or busy with a 0 or 1, respectively. The table called KTO records the line to which a call is being made. Another table, called KEND, records the time at which the call ends. These last two items of data about a call will be kept at the location corresponding to the calling line only.

The maximum number of links and the number currently in use are recorded in locations NLNKS and NUSE, respectively. The clock time is recorded in ICLOCK. Four words are set aside to record details of the next call to enter the system. These are IFROM, ITO, ILNGTH, and ITA recording the origin, destination, length, and arrival time of the call, respectively. A bootstrap method will be used to create new calls; that is, at the time the next call enters the system, the arrival time of its successor will be generated (see Sec. 8-3).

Summarizing these definitions we have

LINES(I)	State of the *i*th line
KTO(I)	Destination of a call originated by line *i*
KEND(I)	Time call from line *i* will end
NLNKS	Maximum number of links
NUSE	Number of links in use
ICLOCK	Clock time
IFROM	Origin of next call
ITO	Destination of next call
ILNGTH	Length of next call
ITA	Time of arrival of next call

The method of arranging for table space in FORTRAN is by means of a DIMENSION statement. The names of the tables are given in a list, with the size of the table. Since the model is being programmed to allow for a maximum of 100 lines, the required statement is

DIMENSION LINES(100), KTO(100), KEND(100)

A three-dimension table of the form LINES(I,J,K) could be used, but three tables are being used so that they can be named separately and make the explanation of the program simpler.

A complete listing of the program is given in Fig. 9-11. It will be described with a series of flow charts, one for each of the major sections of the program. Numbers on the flow charts are program statement numbers. At many points, checks are made for errors. An error results in transfer to an error routine which prints an error stop number or comment. The principal error stops are number 7, which indicates an illegal input card, and number 5, which indicates an incorrect use of a RESET card. The other stops are for conditions that should not occur with normal functioning of the program. Should they occur, a system dump is printed.

9-5
Random Number Subroutine

The program will use both integer and floating-point random numbers, so the generator RAND, described in Sec. 6-7, will be used as a subroutine. It must be supplied with a seed in IX, and following each execution of the routine, the output must be assigned as the next input. Only one series of random numbers is needed, so the statement for assigning the output to be the new input (IX = IY) will be included within the subroutine. The output IY, or IX, is an integer random number between 0 and $2^{31} - 1$, and YFL is a floating-point number between 0 and 1.

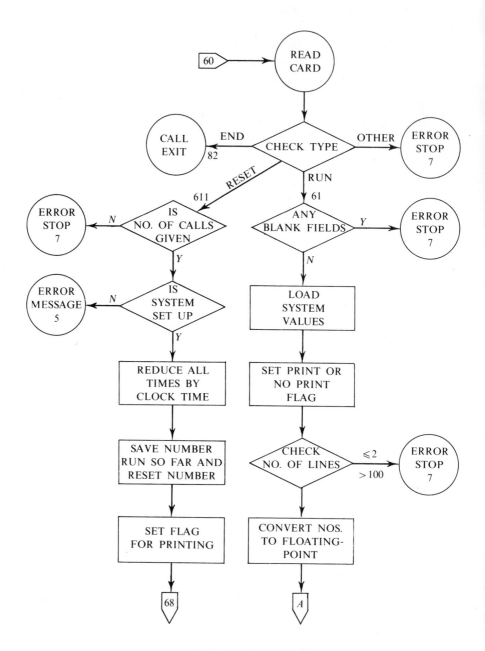

Figure 9-2. Read a card.

The program begins by reading an input card and transferring to one of several locations according to card type. An illegal card number causes error stop 7. Figure 9-2 is a flow chart of the actions taken in reading and checking a card.

Should the card be an END card, the program simply goes to call an EXIT routine at 82. For a RUN card, the first six fields are checked to ensure that they are not negative or zero. If the card is correct, its information is read into a number of locations for future reference. The locations are defined as follows:

NLINES	Number of lines
NLNKS	Number of links
NARR	Mean call inter-arrival time
NLNGTH	Mean call length
NRUN	Number of calls to be run

The seventh field is stored in NPRNT, which will be checked at the end of the run to determine if printing is to be suppressed. The number of lines is then checked to ensure that there are at least 2 lines and that the limit of 100 lines, which results from the fixed table size, is not exceeded. Finally, three of the input numbers are converted to floating-point form for use in the routine creating calls. The floating-point numbers are

ALINES	Number of lines
AARR	Mean call inter-arrival time
ALNGTH	Mean call length

The reading of a RUN card signals the start of a run and the program therefore goes to a section headed "Initialize Program" for which a flow chart is given in Fig. 9-3.

The call tables are first set to zero. Locations NSTR, NFIN, and NUSE are then set to zero. The first is used to save the number of calls run prior to a RESET card; the second is used when scanning the calls in progress. Initially, the program will be empty of calls and it will sometimes happen during the running of the system that there are no calls in progress. To indicate this condition a location called KTFIN, which normally holds the time the next call will finish, is set to the largest value it can take. The number is entered as INF, which is $2^{31} - 1$, or all binary ones in a 32 bit word machine. For computers with other word sizes, it should be redefined. The random number generator seed is then set. Note that this same seed is set at the beginning of every run; so that, if a run is repeated, exactly the same series of random numbers will be produced.

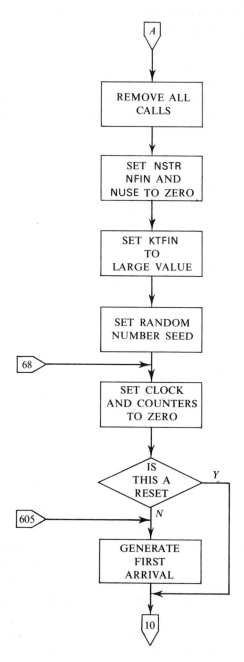

Figure 9-3. Initialize program.

The clock time and the counters for collecting the program output statistics are set to zero. Finally, the first arrival time is generated. The inter-arrival time is assumed to be exponentially distributed and the program uses the method described in Sec. 7-4.

Returning to the flow chart of Fig. 9-2, it is seen if the card is a RESET card, the program checks that the number of calls to be run has been properly defined. It makes sure that the system can be reset by checking that NLINES has been established by a previous RUN card; if not, an error message is printed. All times in the system are then reduced by the clock time at the end of the previous run. The number of calls run up to the end of the previous run is saved in NSTR for later printing. The number of calls now to be run is put in NRUN. and the flag NPRNT, which may have been set to 1 to stop printing from the previous run, is reset to 0. The program will then go to statement 68, where it enters a section also used by a RUN card to set the clock and output statistics counters to 0. The section leads to the generation of the first arrival time, which is not needed when resetting the run, so that action is bypassed for a RESET card. Following either a RUN or RESET card, the program leaves the initialization section by going to statement 10 to begin the main cycle of events.

9-7

Finding the Next Potential Event

There are two types of events in the system; either a call in progress finishes or a new call arrives. The time the next call arrives is in location ITA. The next call to finish can be positioned anywhere in the calls-in-progress tables. The program scans the table KEND and maintains the following records:

| KTFIN | Time at which next call in progress finishes |
| NFIN | Location in the table of the next call to finish |

The way these records are updated will be described later.

A flow chart for the section concerned with finding the next event is shown in Fig. 9-4. If there are no calls in the system, KTFIN will be set to INF. The program will immediately take the next arrival as the next event. If the system is not empty, KTFIN is compared to ITA, the time of the next arrival, and the program updates the clock to the time that is smaller. If they should be equal, a request for a connect and a disconnect of a call appear simultaneously. The program has been written to process the disconnect first. When the disconnect is completed, the program will return to make the connect as a later event. At that time, the program goes through the action of updating the clock, but the value of the clock time will remain the same. It is also possible to have

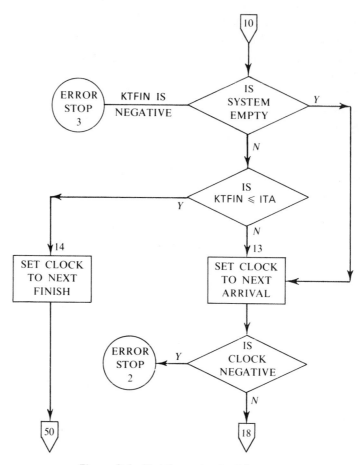

Figure 9-4. Find the next potential event.

simultaneous events caused by more than one disconnect occurring at the same time, or more than one call arriving simultaneously. If the next event is that a call terminates, the program goes to statement 50 to disconnect the call.

When there is a new arrival, the details of the new call, other than its arrival time, must be generated at that time because the conditions existing in the system determine the possible origins for the call. The program goes to statement 18 to execute the portion of the program, concerned with creating a call, for which a flow chart is given in Fig. 9-5.

It will be assumed that a new arrival can come, with equal probability, from any line not currently busy and that it is equally likely to go to any line other than itself irrespective of whether it is busy or not. The program picks a line at random as a possible origin and then checks whether that line is busy. If it is, the program makes another choice. This is a simple procedure but it is time-consuming if the system is heavily loaded. In fact, if all lines are busy, it will

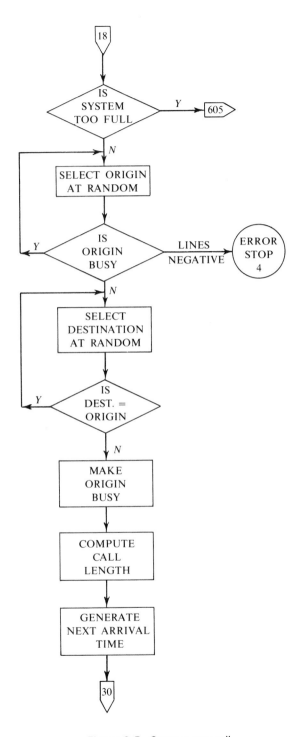

Figure 9-5. Create a new call.

result in an endless loop, so a check is made first to see that at least two lines are open before proceeding to pick the origin. If not, the program abandons the call and returns to the initialization routine to create a new arrival at a later time. The destination is also picked at random and a new choice made if it should happen to be the same as the origin.

The method used to pick a line at random is the one described in Sec. 6-8. The number of lines is multiplied by a random number, and the integral part of the result is increased by 1, so that the integers 1 through NLINES are equally likely to occur.

Having chosen the origin and destination and stored the results in IFROM and ITO, respectively, the program makes the origin busy and creates the call length. This is assumed to be exponentially distributed. Finally, the program creates the arrival time of the next call and then goes to statement 30 to see if the call that just arrived can be connected.

9-8

Testing if a Call Can be Connected

The flow chart for the portion of the program concerned with checking whether a call can be connected is shown in Fig. 9-6. The program checks if a link is available by seeing if the number in use equals the total number of links. If a link is available, the program checks whether the called party is busy. Since a busy condition is indicated by setting the value of LINES to 1, and the called party line number is in ITO, the check is to see if LINES(ITO) is 1. There are

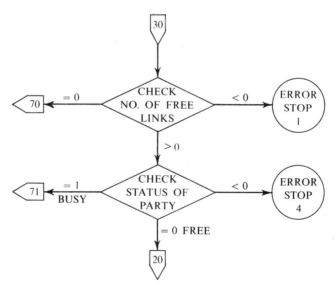

Figure 9-6. Test if a call can be made.

three possible exits from the routine. They are to statement

> 70 If all links are in use
> 71 If the called party is busy
> 20 If the call can be connected

A flow chart for the portion of the program connecting a call is shown in Fig. 9-7. The LINES table is updated to show that the called line has become busy by placing a 1 in the corresponding location of the table. The number of links in use, NUSE, is increased by 1 and the calls-in-progress table is updated by entering the number of the called line in table KTO and the time the call will finish in table KEND. It will be recalled that the convention is to make these entries at the location of the calling number, while the entries at the called number are left zero. The time the call finishes is derived by adding its length ILNGTH to the time of arrival ITA.

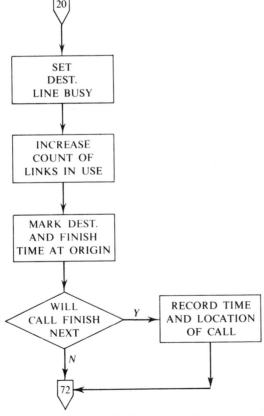

Figure 9-7. Connect a call.

Records are kept in KTFIN and NFIN of the time the next call will finish and its location in the call tables. It is possible that the call just connected is the next to finish; if so, the records are updated. The connection of a call completes an event and the program should return to statement 10 to begin the next event. As shown in the listing, the program actually goes to statement 72 which sends the program to 10. This step is introduced to prepare for an expansion of the program in the next chapter.

9-10

Disconnecting a Call

Figure 9-8 shows a flow chart for the disconnection of a call. The entries in the LINES table for the origin and destination of the call are both set to zero. The number of the originating line is in NFIN. The number of the called line is in the KTO table, located at the position of the originating line. The called number is extracted from KTO and placed in a temporary location NTEMP which is used to set the called line to zero. The fact that a link has now become free is recorded.

When a call is terminated, the records of the next call to finish are no longer valid, so the program must update them. It does so by scanning the table KEND, which records the times the calls in progress will finish. If there are no calls in progress, there are also no links in use, so the value of NUSE is checked. If it is zero, the program omits this section of program and goes to statement 73. For the lines to which calls are being made, and for lines that do not have a call, the entry in KEND is zero, so the search must find the least non-zero value in KEND. The scan ranges from 1 to NLINES, the total number of lines. It is initialized by setting KTFIN to INF, and proceeds by comparing each non-zero value in KEND with KTFIN. The first non-zero value replaces INF in KTFIN, and the location of the line making the call goes to NFIN. Subsequent comparisons determine if other non-zero values are less and, if so, update the records.

9-11

Gathering Statistics

The program is to gather a number of statistics, and the following locations are defined for that purpose:

NBLKS	Count of blocked calls
NBUSY	Count of busy calls
NCOMP	Count of completed calls
NFINS	Count of calls finished

The quantity NFINS is defined as the sum of all calls that the system has processed; it includes completed calls, blocked calls, and busy calls. When NFINS reaches the total of NRUN, which was read from the input card, the simulation run ends.

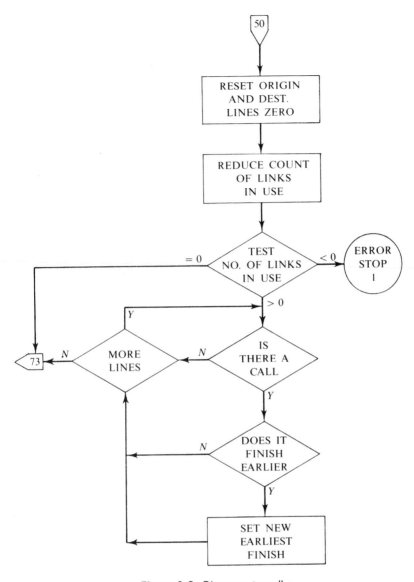

Figure 9-8. Disconnect a call.

The block diagram for the gathering of statistics is shown in Fig. 9-9. It can be entered at four points, for the following reasons:

70	Call was blocked
71	Called line was busy
72	Call was connected
73	Call was disconnected

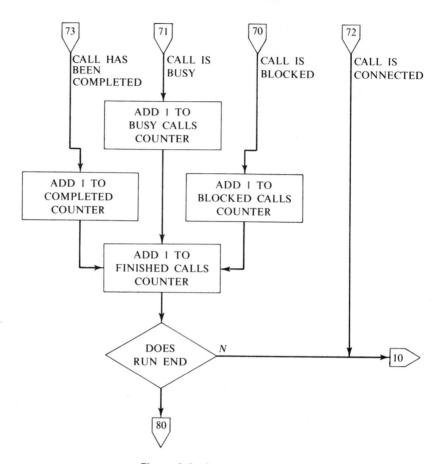

Figure 9-9. Gathering statistics.

When a call is completed, the counter **NCOMP** is incremented by 1. For busy or blocked calls, the appropriate counter is incremented by 1; for busy, blocked, or disconnected calls, the counter **NFINS** is incremented. No counter needs to be changed if a call is connected.

If the number of finished calls equals the number established for the simulation run, the program goes to statement 80 to terminate the simulation. Otherwise, the program returns to statement 10 to execute the next event.

9-12

Output
Report

When a run is started by a **RUN** card with a non-zero field 7, **NPRNT** is non-zero. If the program finds this condition, it bypasses the output report section and transfers to read the next input card. Otherwise, an output report is produced by the last section of the program. This section will not be described

```
NO. OF LINES                    5C
NO. OF LINKS                    1C
MEAN CALL INTERVAL              12
MEAN CALL LENGTH                12C
NUMBER OF CALLS                 10CC


RUN RESET AFTER    1C  CALLS FINISHED


CLOCK TIME                   11131

CALLS COMPLETED               661
BUSY CALLS                    235
BLOCKED CALLS                 1C4
```

Figure 9-10. Program output report.

in full since it is mainly concerned with the preparation of output messages. It is, however, included in the listing of Fig. 9-11. A typical output produced by the routine is shown in Fig. 9-10.

```
C        SIMULATION OF TELEPHONE SYSTEM - MODEL 1
C
C        PROGRAM REQUIRES INPUT CARDS WITH THE FOLLOWING INFORMATION
C        FIELD NO.   COLUMNS    INFORMATION
C
C        RUN CARD
C
C           1       1 -  6     CARD TYPE = 1
C           2       7 - 12     NUMBER OF TELEPHONE LINES
C           3      13 - 18     NUMBER OF LINKS
C           4      19 - 24     MEAN CALL INTER-ARRIVAL TIME
C           5      25 - 30     MEAN CALL LENGTH
C           6      31 - 36     NUMBER OF CALLS TO BE RUN
C           7      37 - 42     PRINT CONTROL  0 = PRINT
C                                             1 = NO PRINT
C
C
C        RESET CARD
C
C           1       1 -  6     CARD TYPE = 3
C           6      31 - 36     NUMBER OF CALLS TO BE RUN
C
C        END CARD
C
C           1       1 -  6     CARD TYPE = 4
C
C        LOCATIONS IN THE PROGRAM HAVE THE FOLLOWING MEANINGS
C
C           IFROM  NEXT FUTURE CALL ORIGIN
C           ILNGTH NEXT FUTURE CALL LENGTH (IF PLACED)
C           ITA    NEXT FUTURE CALL ARRIVAL TIME
C           ITO    NEXT FUTURE CALL DESTINATION
C           KTFIN  TIME NEXT CALL IN PROGRESS FINISHES
C           NARR   MEAN INTER-ARRIVAL TIME
C           NBLKS  NO OF BLOCKED CALLS
C           NBUSY  NO OF BUSY CALLS
C           NCOMP  NO OF COMPLETED CALLS
C           NERR   ERROR STOP LOCATION NUMBER
C           NFINS  NO OF CALLS PROCESSED
C           NLINES NO OF LINES
C           NLNGTH MEAN CALL LENGTH
C           NRUN   NO OF CALLS TO BE RUN
C           NUSE   NUMBER OF LINKS IN USE
C
C        GENERATE A RANDOM NUMBER - YFL
C
0001             SUBROUTINE RAND (IX,IY,YFL)
C
0002             IY = IX*1220703125
0003             IF (IY) 1,2,2
0004      1      IY = IY + 2147483647 + 1
0005      2      YFL = IY
0006             YFL = YFL*.4656613E-9
0007             IX = IY
0008             RETURN
0009             END

C        DEFINITION OF THE SYSTEM IMAGE
C
C           TABLE CALLED LINES HAS ONE WORD PER
C           LINE. VALUE IS 0/1 IF LINE IS FREE/BUSY
C           MINIMUM SIZE IS 2, MAXIMUM SIZE IS 100
C
C           TABLES CALLED KTO AND KEND ARE FOR
C           CALLS IN PROGRESS. EACH ACTIVE CALL
C           IS PLACED AT THE LOCATION INDICATED
C           BY THE ORIGINATING LINE.
C
C           KTO    IS FOR THE CALLED PARTY
C           KEND   IS TIME THE CALL WILL FINISH
C
0001             DIMENSION LINES(100),KTO(100),KEND(100),ICARD(7)
C
C           INF IS THE LARGEST VALUE A VARIABLE CAN TAKE.
C           KTFIN IS SET TO INF TO INDICATE AN EMPTY SYSTEM
C
```

Figure 9-11. Listing of telephone system simulation in FORTRAN-model 1.

```
        C       READ AND CHECK A CARD
        C
0002     60            READ (5,1,ERR = 107) (ICARD(I),I = 1,7)
0003      1            FORMAT (7I6)
0004                   ITEM = ICARD(1)
0005                   GO TO (61,107,611,82), ITEM
0006                   GO TO 107
        C
        C              CHECK RESET CARD
        C
0007    611            IF (ICARD(6)) 107,107,612
0008    612            IF (NLINES) 105,105,613
0009    613            DO 615 I = 1,NLINES
0010                   IF (KEND(I)) 105,615,614
0011    614            KEND(I) = KEND(I) - ICLOCK
0012    615            CONTINUE
0013                   ITA = ITA - ICLOCK
0014                   IF (INF - KTFIN) 618,618,617
0015    617            KTFIN = KTFIN - ICLOCK
0016    618            NSTR = NRUN
0017                   NRUN = ICARD(6)
0018                   NPRNT = 0
0019                   GO TO 68
        C
        C              CHECK RUN CARD
        C
0020     61            DO 64 I = 2,6
0021                   IF (ICARD(I)) 107,107,64
0022     64            CONTINUE
0023                   NLINES = ICARD(2)
0024                   NLNKS = ICARD(3)
0025                   NARR = ICARD(4)
0026                   NLNGTH = ICARD(5)
0027                   NRUN = ICARD(6)
0028                   NPRNT = ICARD(7)
0029                   IF (NLINES - 2) 107,107,62
0030     62            IF (100 - NLINES) 107,67,67
        C
        C              CONVERT TO FLOATING POINT
        C
0031     67            ALINES = NLINES
0032                   AARR = NARR
0033                   ALNGTH = NLNGTH
        C
        C       INITIALIZE PROGRAM
        C
0034                   DO 66 I = 1,NLINES
0035                   LINES(I) = 0
0036                   KTC(I) = 0
0037     66            KEND(I) = 0
0038                   NSTR = 0
0039                   NUSE = 0
0040                   NFIN = 0
0041                   INF = 2147483647
0042                   KTFIN = INF
0043                   IX = 13579
        C
        C              ENTER HERE FOR RESTART
        C
0044     68            ICLOCK = 0
0045                   NBLKS = 0
0046                   NBUSY = 0
0047                   NCCMP = 0
0048                   NFINS = 0
        C
        C              COMPUTE FIRST ARRIVAL
        C
0049                   IF (ICARD(1) - 3) 605,10,605
0050    605            CALL RAND (IX,IY,YFL)
0051                   ITA = -AARR*ALOG(YFL) + ICLOCK
0052                   GO TO 10
```

Figure 9-11. Listing of telephone system simulation in FORTRAN-model 1 (*cont.*).

```
                    C
                    C
                    C          FIND NEXT POTENTIAL EVENT
0053        10              IF (KTFIN - INF) 11,13,103
0054        11              IF (KTFIN - ITA) 14,14,13
0055        13              ICLOCK = ITA
0056                        IF (ICLOCK) 102,18,18
0057        14              ICLOCK = KTFIN
CC58                        GO TO 50
                    C
                    C              CREATE A NEW CALL
                    C
0059        18              IF (NLINES - 2*NUSE - 2) 605,605,15
0060        15              CALL RAND (IX,IY,YFL)
0061                        IFROM = ALINES*YFL + 1
0062                        IF (LINES(IFROM)) 104,16,15
0063        16              CALL RAND (IX,IY,YFL)
CC64                        ITO = ALINES*YFL + 1
0065                        IF (IFROM - ITO) 17,16,17
0066        17              CALL RAND (IX,IY,YFL)
0067                        ILNGTH = -ALNGTH*ALOG(YFL)
C068                        LINES(IFROM) = 1
                    C
                    C              COMPUTE NEXT ARRIVAL
                    C
0069                        CALL RAND (IX,IY,YFL)
CC70                        ITA = -AARR*ALOG(YFL) + ICLOCK
0071                        GO TO 30
                    C
                    C      DISCONNECT A CALL
                    C
                    C              NFIN HAS LOCATION OF CALL
0072        50              LINES(NFIN) = 0
0073                        NTEMP = KTO(NFIN)
0074                        LINES(NTEMP) = 0
0075                        NUSE = NUSE - 1
0076                        KTO(NFIN) = 0
CC77                        KEND(NFIN) = 0
                    C
                    C              UPDATE NEXT-TO-FINISH RECORDS
                    C
0078        51              KTFIN = INF
0079                        IF (NUSE) 101,73,52
0080        52              DO 55 I = 1,NLINES
CC81                        IF (KEND(I)) 55,55,53
0082        53              IF (KEND(I) - KTFIN) 54,55,55
C083        54              KTFIN = KEND(I)
0084                        NFIN = I
0085        55              CONTINUE
0086                        GO TO 73
                    C
                    C      TEST IF CALL CAN BE MADE
                    C
C087        30              IF (NLNKS - NUSE) 101,70,31
0088        31              IF (LINES(ITO)) 104,20,71
                    C
                    C      CONNECT A CALL
                    C
0089        20              LINES(ITO) = 1
0090                        NUSE = NUSE + 1
0091                        KTO(IFROM) = ITO
0092                        KEND(IFROM) = ICLOCK + ILNGTH
0093                        IF (KEND(IFROM) - KTFIN) 21,72,72
0094        21              KTFIN = KEND(IFROM)
0095                        NFIN = IFROM
0096                        GO TO 72
                    C
```

Figure 9-11. Listing of telephone system simulation in FORTRAN-model 1 (*cont.*).

```
        C       GATHER STATISTICS
        C
CC97    70      NBLKS = NBLKS + 1
0098            LINES(IFROM) = 0
CC99            GC TO 74
0100    71      NBUSY = NBUSY + 1
0101            LINES(IFROM) = 0
0102            GC TO 74
0103    72      GO TO 10
0104    73      NCOMP = NCOMP + 1
0105    74      NFINS = NFINS + 1
0106            IF (NRUN - NFINS) 80,80,10
        C
        C       COMPUTE FINAL STATISTICS AND PRINT OUT
        C
0107    80      IF (NPRNT) 6C,801,6C
01C8    801     WRITE(6,7)NLINES,NLNKS,NARR,NLNGTH,NRUN
0109    7       FORMAT(1H1,9X,12HNO. OF LINES,12X,I6/10X,12HNO. OF LINKS,
                112X,I6/1CX,18HMEAN CALL INTERVAL,7X,I5/
                210X,16HMEAN CALL LENGTH,9X,I5/10X,15HNUMBER OF CALLS,9X,I6///)
C110            IF(NSTR)105,803,802
0111    8C2     WRITE(6,95)NSTR
C112    95      FORMAT(1CX,15HRUN RESET AFTER,I5,16H  CALLS FINISHED//)
0113    803     WRITE(6,2)ICLOCK,NCOMP,NBUSY,NBLKS
0114    2       FORMAT(1CX,10HCLCCK TIME,11X,I9//10X,15HCALLS COMPLETED,
                16X,I9/1CX,10HBUSY CALLS,11X,I9/10X,13HBLOCKED CALLS,8X,I9)
0115            GO TO 6C
        C
C116    82      CALL EXIT
        C
        C       ERROR ROUTINE
        C
0117    101     NERR = 1
C11E            GO TO 90
C119    102     NERR = 2
012C            GC TC 90
C121    103     NERR = 3
0122            GO TO 90
0123    104     NERR = 4
0124    90      WRITE (6,8) NERR
0125    8       FORMAT(//10X,13HERRCR STOP NO,5X,I5//)
012E    91      WRITE (6,4) NUSE, KTFIN, IFROM, ITC, ITA
C127    4       FORMAT(5I6//)
C128            DO 92 I = 1,NLINES
C129            WRITE (6,5) LINES(I), KTO(I), KEND(I)
C13C    5       FORMAT(5X,4(5X,I6))
0131    92      CONTINUE
0132            GC TC 80
C133    105     WRITE (6,96)
C134    96      FORMAT(//1CX,18HILLEGAL RESET CARD//)
C135            GO TO 6C
C136    107     WRITE (6,93)
C137    93 ·    FCRMAT(//1CX,16HINPUT CARD ERROR//)
C13E            GO TC 6C
C139            END
```

Figure 9-11. Listing of telephone system simulation in FORTRAN - model 1 (*cont.*).

Exercises   9-1 Use the telephone simulation program to investigate how the number of blocked and busy calls vary as the number of links is changed from 4 to 16 in steps of 2. Keep all other parameters unchanged.

9-2 Change the FORTRAN program for the telephone system so that the call lengths are normally distributed. Keep the mean call length of 120 and assume that the standard deviation is 20.

9-3 Assume calls cost 1¢ a second and modify the telephone system simulation to gather the revenue earned by the telephone company.

9-4 Assume that 10% of the calls made in the telephone system are for time or weather information. These calls do not need a link and they can always be connected. They take 20 ± 10 seconds to be completed. Modify the program to include these calls.

9-5 Modify the telephone system simulation to allow for the fact that it takes 5 ± 3 seconds to connect the calls or find out that the called line is busy. A link is being held while the connection or the test is being made.

9-6 Draw a flow chart showing how to write a simulation program for a gas filling station. Assume each car served needs an attendant and a pump and that, if a car cannot get immediate service, it leaves.

9-7 Rewrite the routine for assigning origins in the telephone system assuming that there are 50 lines but the probability that a call will come from lines 1 to 10 is twice the probability of its coming from the other lines.

Bibliography   1 Murphy, J. G., "A Comparison of the Use of GPSS and SIMSCRIPT Simulation Languages in Designing Communications Networks," Tech. Memo. TN-03969, MITRE Corp., Bedford, Mass., 1964.

2 Young, Karen, "A User's Experience With Three Simulation Languages (GPSS, SIMSCRIPT, and SIMPAC)," TM-1755/000/00, Systems Development Corp., Santa Monica, Calif., 1963.

3 Weinert, Arla E., "A SIMSCRIPT–FORTRAN Case Study," *Comm. ACM*, X, No. 12 (Dec., 1967), 784–792.

10

SIMULATION PROGRAMMING TECHNIQUES

The exact statistics required from a model depend upon the study being performed, but there are certain commonly required statistics for which programming techniques have been developed. Among the commonly needed statistics are

(a) *Counts* giving the number of entities of a particular type or the number of times some event occurred.

(b) *Summary statistics*, such as extreme values, mean values, and standard deviations.

(c) *Utilization*, defined as the fraction (or percentage) of time some entity is engaged.

(d) *Occupancy*, defined as the fraction (or percentage) of a group of entities in use on the average.

(e) *Distributions* of important variables, such as queue lengths or waiting times.

(f) *Transit times*, defined as the time taken for an entity to move from one part of the system to some other part.

When there are stochastic effects operating in the system, all these system measures will fluctuate as a simulation proceeds, and the particular values reached at the end of the simulation are taken as estimates of the true values they are designed to measure. Deciding upon the accuracy of the estimates is a difficult problem that will be taken up in Chap. 15. For the time being, we discuss the methods used to derive the estimates.

10-2

Counters and Summary Statistics

Counters are the basis for most statistics. Some are used to accumulate totals; others record current values of some level in the system. The telephone system simulation of the last chapter, for example, used counters to record the total number of lost and busy calls, as well as to keep track of how many links were in use at any time. Maxima or minima are easily obtained. Whenever a new value of a count is established, it is compared with the record of the current maximum or minimum, and the record is changed when necessary.

The mean of a set of N observations x_i $(i = 1, 2, \ldots, n)$ is defined as

$$M = \frac{1}{n} \sum_{i=1}^{n} x_i$$

A mean, therefore, is derived by accumulating the total value of the observations, and also accumulating a count of the number of observations. The division of the two numbers is performed at the end of the simulation run in preparation for the final output of the program.

A standard deviation, S, is derived from the square root of the sum of the squares of the deviations of individual observations from the mean.

$$S = \left\{ \frac{1}{(n-1)} \sum_{i=1}^{n} (M - x_i)^2 \right\}^{1/2}$$

The common method of summing the squares is based on the following expansion:

$$\sum_{i=1}^{n} (M - x_i)^2 = \sum_{i=1}^{n} x_i^2 - nM^2$$

The accumulated sum of the observations must be kept to derive the mean value, so the only additional record needed to derive a standard deviation is the sum of the squares. It can take large values, especially when many observations are made. Then, double precision calculations may be needed, particularly if the calculations are carried out in integer arithmetic. For a discussion of methods for calculating means and standard deviations when extremely high precision is needed, see Ref. (1).

A common requirement of a simulation is measuring the load on some entity, such as an item of equipment. The simplest measure is to determine what fraction of the time the item was engaged during the simulation run. The term *utilization* will be used to describe this statistic. Typically, the time history of the equipment usage might appear as shown in Fig. 10-1. To measure

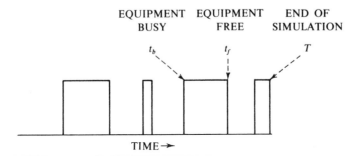

Figure 10-1. Utilization of equipment.

the utilization, it is necessary to keep a record of the time t_b at which the item last became busy. When the item becomes free at time t_f, the interval $t_f - t_b$ is derived and added to a counter. At the end of the simulation run, the utilization U is derived by dividing the accumulated total by the total time T, so that:

$$U = \frac{1}{T} \sum_{1}^{N} (t_f - t_b)$$

A discrete simulation program, updating time as events occur, will measure the intervals $t_f - t_b$ directly. A continuous simulation program updating time in small intervals will need to build up the count by counting the number of intervals in which the item is busy.

Note that it is important to check whether the item is busy at the end of the run, and, if so, add to the counter a quantity representing the engagement from the last time it became busy to the end of the simulation. Correspondingly, it is also important to check the initial conditions to see if the entity is busy at the beginning of the run.

In dealing with groups of entities, rather than individual items, the calculation is similar, requiring that information about the number of entities involved also be kept. Figure 10-2, for example, represents as a function of time the number of links in a telephone system that are busy. To find the average number of links in use, a record must be kept of the number of links in use and the time at which the last change occurred. If the number changes at time t_i to the value n_i, then, at the time of the next change t_{i+1}, the quantity $n_i(t_{i+1} - t_i)$ must be calculated and added to an accumulated total. The average number in use

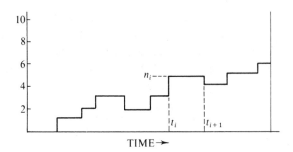

Figure 10-2. Time history of busy telephone links.

during the simulation run, A, is then calculated at the end of the run by dividing the total by the total simulation time T, so that:

$$A = \frac{1}{T} \sum_{i=1}^{N} n_i(t_{i+1} - t_i)$$

Figure 10-2 might also represent the number of entities waiting on a queue, in which case, the calculation gives the mean number of entities waiting.

If there is an upper limit on the number of entities, as there is a limit on the number of links in a telephone system, the term *occupancy* is often used to describe the average number in use as a ratio to the maximum. Thus, if there are N links in a telephone exchange and the quantity n_i is the number busy in the interval t_i to t_{i+1}, the average occupancy is

$$B = \frac{1}{NT} \sum_{i=1}^{T} n_i(t_{i+1} - t_i)$$

As mentioned before, it is important to check the conditions at the beginning and end of the simulation.

10-4

Recording Distributions

Determining the distribution of a variable requires counting how many times the value of the variable falls within specific intervals. A table sets aside locations in which to record the values defining the intervals and to accumulate each count. As each new observation is made, its value is compared with the limits established for the intervals, and 1 is added to the counter for one interval. As a rule, the intervals are of uniform size.

The nature of the random variable being measured determines when the observations are made. If the mean waiting time for a service were to be

Simulation Programming Techniques   /   Ch. 10

measured, an observation would be taken as each entity starts to receive service, so that the times at which the observations are tabulated, or recorded for tabulation, are randomly spaced. To measure the distribution of the number of entities waiting, however, observations would be taken at uniform intervals of time.

It is customary to accumulate the number of observations and the sum of the squares to calculate the mean value and the standard deviation at the same time the distribution is being derived. Each observation, x_i, will therefore result in a count of 1 being added to the appropriate counter, the addition of x_i to the accumulated total $\Sigma\, x_i$ and the addition of x_i^2 to the sum $\Sigma\, x_i^2$. Note that, while the distribution derived is an approximation, since it matches the values of the observations to an interval, the mean or standard deviation will be accurate within the accuracy limits of the computer, even if some observations fall outside the table limits.

In addition to deriving the mean and standard deviation of a distribution, the final output will often express the data in some other convenient form. For example, the cumulative distribution may be given, or the distribution may be rescaled to express the counts as percentages of total observations or express the interval size in terms of the mean value of the observations. All these derived results can be calculated at the end of the run. A typical example of such a table print-out will be found in Fig. 11-11.

To measure transit times, the clock is used in the manner of a time stamp. When an entity reaches a point from which a measurement of transit time is to start, a note of the time of arrival is made. Later, when the entity reaches the point at which the measurement ends, a note of the clock time upon arrival is made and compared with the first time to derive the elapsed interval.

10-5

FORTRAN Programs for Statistics

To illustrate the techniques discussed in the last few sections, we now write a few simple programs that gather statistics in the FORTRAN program for the telephone system given in the last chapter.

Suppose first, that statistics are gathered on the occupancy of the links. Allowance was previously made for recording the maximum number of links and the number of links currently in use. We add an extra location to record the time at which the number of links in use changed and a location to hold the accumulation of occupancy counts. We have

NLINKS	Maximum number of links
NUSE	Number of links in use
NLAST	Last time at which NUSE changed
NSUM	Accumulated links occupancy count

Referring to Fig. 9-9, which describes the gathering of statistics, we see the program transfers to statement number 72 when a call is connected and to statement number 73 when a call is disconnected. At both points, two new statements will be added to record the statistics needed to calculate the occupancy of the links. They compute the interval since the last change by subtracting NLAST from ICLOCK and multiply the result by the number of links that were used during the interval. NLAST is then updated.

Suppose next that callers finding all links busy do not hang up, but wait for a link to become available. The simulation is to find the mean time they must wait and the distribution of the waiting time. The modifications that must be made to arrange for delaying calls will be discussed in later sections. For the time being, assuming that the time a caller has had to wait is available, we show here a program for tabulating the waiting time.

Normally, the tabulation intervals are uniform in size and the specifications will be for

(a) The lower limit of tabulation
(b) The interval size
(c) The number of intervals

The meaning of these terms is illustrated in Fig. 10-3. The user will not always gauge accurately what the potential range of values will be, so it is

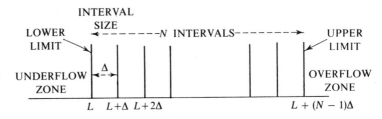

Figure 10-3. Definition of a distribution table.

customary to count how many times an observation falls below the lower limit and beyond the upper limit.

Suppose the table is defined by a TABLE card with the following format:

Field	Columns	Information
1	1– 6	Card type = 2
2	7–12	Lower limit
3	13–18	Interval size
4	19–24	Number of intervals

The routine for reading input cards and initializing the program must be modified to read the card, check its validity, and establish the space required for recording the statistics. Details of the programming are given in the final listing shown in Fig. 10-10. An arbitrary maximum limit of 100 tabulation intervals will be made. In computing the number of intervals, the intervals for observations falling below the lower limit and beyond the upper limit are to be included in the count of the intervals, so that the number of intervals should be at least three. The program is arranged so that once a TABLE card has been loaded, it remains in effect for all subsequent runs. It can, however, be superseded by a new TABLE card, or it can be cancelled by a TABLE card with the number of intervals *and* the interval size set at zero.

To carry a table from one run to the next, each new RUN card clears the data gathered in the table from the previous run. To this end, the number of intervals is stored in a location called NINTS and used as a flag to indicate that a table has been set up. Upon finding the flag non-zero, the program clears the table.

Space for the table is set aside in an array ITAB(I,J), whose size is declared in a DIMENSION statement to be 100×2. The space is arranged as shown in Fig. 10-4. Three separate locations are also added to act as counters for deriving summary statistics for the tabulated data. These are:

TTOL   The total number of observations made
TSUM   The sum of the observations
TSSQ   The sum of the squares of the observations

It will be noticed that the interval sizes and the count of the observations are being kept as integral numbers, while the totals for the summary statistics are kept as floating-point numbers.

The routine for carrying out the tabulation assumes that the quantity to be tabulated is placed in a location called INPT. It locates the appropriate interval to which the observation belongs by comparing the observation against the interval limits, and it adds 1 to the corresponding counter. It then increments the counters for the summary statistics. Coding for the routine is shown in the listing of Fig. 10-10.

In a larger program requiring several tables, both the table loading routine and the tabulation routine would be generalized to allow for several tables, each identified by its own number. In that case, the tabulation routine would be organized as a subroutine to be executed wherever it is needed. In the present program, the routine will be entered whenever a delayed call is finally connected.

	$J = 1$	$J = 2$
$I = 1$	L LOWER LIMIT	UNDERFLOW COUNT
	$L + \Delta$	1st INTERVAL COUNT
	$L + i\Delta$	ith INTERVAL COUNT
$I = N - 1$	$L + (N - 1)\Delta$	$(N - 1)$th INTERVAL COUNT
$I = N$		OVERFLOW COUNT

TTOL	NUMBER OF ENTRIES
TSUM	TOTAL OF ENTRIES
TSSQ	SUM OF SQUARES

Figure 10-4. Space reserved for a distribution table.

10-6

List Structures

It is frequently required in a simulation program to keep a set of records in a particular order where the number in the set fluctuates. For example, when callers who find all links busy wait for one to become free, the number of waiting calls will fluctuate and, assuming a first-come, first-served queuing discipline, they must be kept in order of arrival.

An effective way of maintaining such records is by using a *list structure*. The records are then said to be in a *list* or on a chain. The records of the entities in the list are identified in the computer memory by an address. If the records have more than one word, the address is standardized to one of the words, such as the first. One word, or field in a word, called a *pointer*, is set aside in each record for the purpose of structuring the list. In addition, a special word called the *list header* is provided for entering the list.

The records are chained together as illustrated in Fig. 10-5. The list header contains the address of the first record in the list. The pointer of the first record contains the address of the second record, and so on up the list. The last record

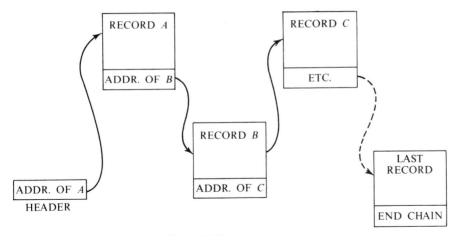

Figure 10-5. List structure.

in the list contains a special end-of-chain symbol in the pointer space to indicate that it is the last member. If the list happens to be empty, the list header usually contains the end-of-chain symbol.

Beginning from the header, the program is able to move up the list by following the chain of pointers. If the program needs to remove a record from the list, say record *B* from the list *ABC*..., it simply changes the pointer in *A* to point to *C*, as illustrated in Fig. 10-6. Correspondingly, to insert a record into the list, for example, to put *Z* in the list *ABC*... between *B* and *C*, the pointer *B* is set to *Z* and the pointer of *Z* is set to *C*.

With a first-in, first-out rule of ordering the records, it is convenient also to keep a *list trailer* that has the address of the last record, because new additions are made at the end. The trailer record avoids the necessity of working along

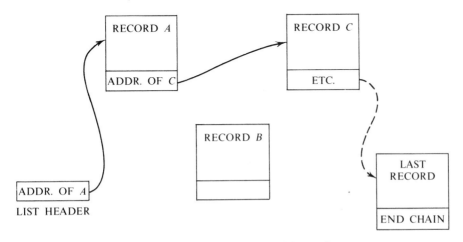

Figure 10-6. Removing a record from a list.

the list to find the last entry. The trailer will also contain the end-of-chain symbol when the list is empty.

It is sometimes necessary to scan the list from either end, in which case a second pointer is added to the records and the trailer record becomes the start for a search in the reverse direction, as shown in Fig. 10-7. Such lists are called double-threaded or two-way lists. Other more complicated organizations are possible to allow sub-lists to be connected to individual members of a list or to allow one record to be a member of more than one list (2).

To illustrate the use of a list, we modify the FORTRAN program of Chap. 9 to allow calls finding all links busy to wait for one to become available on a first-come, first-served basis. The system image must be expanded by adding two words per line in the calls-in-progress table; one for the pointers and one to

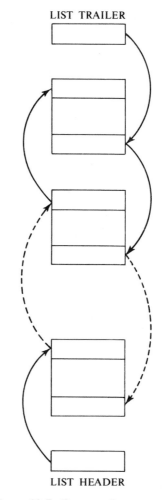

Figure 10-7. Two-way list structure.

preserve the call length. It is also necessary to set aside words for the list header and list trailer. We define:

KPNTR(I)	A table for pointers
KLNGTH(I)	A table for the call length of delayed calls
KHEAD	List header
KTAIL	List trailer

The number INF (which is all binary ones) will be used as an end-of-chain symbol. Following the convention previously established, the records of the blocked calls will be kept at the entry for the originating line. The address of a blocked call is therefore the number of the originating line. The table KTO will still contain the destination of the call. If the call is blocked, the table KEND will be used to record the time the call arrived; so that, when it is finally connected, it is possible to measure how long the call waited. To avoid confusing arrival and finish times, the arrival times will be entered with a *negative* sign.

A block diagram of a routine for adding blocked calls to the end of the list is shown in Fig. 10-8 and the coding is shown in the listing of Fig. 10-10.

As before, a count will be kept of the number of blocked calls in the location NBLKS. In addition, a counter NWAIT is established to keep a count of how many calls are waiting to be connected. Both of these counters are incremented upon entering the routine. If the list is found to be empty, the header is set to the address of the new record. Otherwise, the pointer of the old last record (whose address is to be found in the trailer) is set to the address of the new last record. The pointer of the new record is marked with the end-of-chain symbol and the trailer is updated to point to the new last record. The records in the various tables for the calls in the system are then updated. At the address of the originating line, the destination of the new call is entered in KTO, minus the time of arrival is put in KEND, and the call length is put in KLNGTH. The routine is placed at a point where it is entered after a call is found to be blocked. It begins with statement number 70, which is the point reached when the routine testing for conditions finds that all links are in use.

It is now necessary to check whether the disconnection of a call will allow a blocked call to proceed. The program goes to statement 73 when a call has been disconnected. In Sec. 10-5, a routine was inserted at that point to update the link statistics. From there, the program is sent to a routine that checks for waiting calls and removes the first one when it exists. The procedure followed is shown in Fig. 10-9. The coding is shown in the listing of Fig. 10-10.

If the list is empty, no action is taken. Otherwise, details of the first call on the list are entered in IFROM, ITO, and ILNGTH as though it were the next

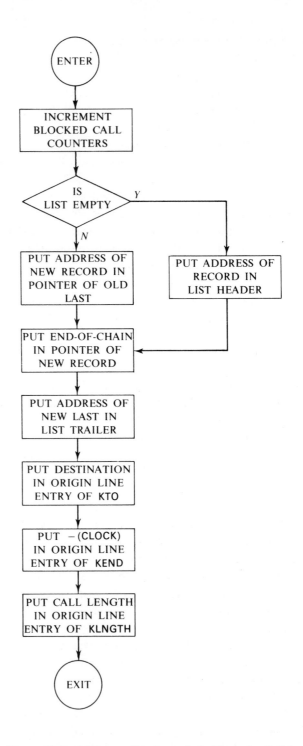

Figure 10-8. Adding a call to the end of a blocked-calls list.

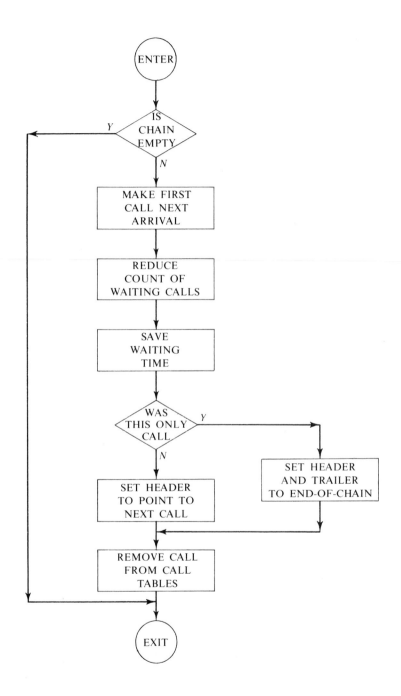

Figure 10-9. Removing a blocked call from the list.

arriving call. The count of waiting calls is reduced and the time spent waiting is calculated and stored in INPT for later tabulation. If the call was the only one on the list, which is tested by seeing if the header points to the same call as the trailer, then both header and trailer are set to end-of-chain to indicate that the list has now been emptied. If there was more than one call on the list, the header is set to point to the next call on the list. Finally, all traces of the call saved in the various call tables are removed.

When this routine has been executed, the program goes to the routine for tabulating waiting times, described in Sec. 10-5. That routine, after executing the tabulation, sends the program to statement 30, which is the point which the program normally reaches when it is about to attempt a connection.

The simple list structure method that has been used in the program is not very efficient in the use of storage space, since space has been set aside for all lines to have a waiting call. More typically, list structures are used not only as a convenient way of linking together a variable number of records, but also as a way of saving space. An efficient list structure would only allocate space for calls when they arrive in the system. Both the calls in progress and the calls waiting to be connected would be listed on separate lists. When a call finishes, the space used for its record would be returned for general use by the program. As can be seen, a list structure does not place any restriction on where the records are physically located, so a list structure program is able to share a limited amount of space between several fluctuating needs. For a more complete discussion of list processing methods, see (3).

10-7

Extension of the Telephone System Example

Bringing together all the modifications that have been developed in this chapter, Fig. 10-10 is a listing of the new version of the program in which blocked calls wait for a link on a first-come, first-served basis. The program now gathers statistics for link occupancy and the distribution of delay time. Figure 10-11 is a typical output. It was produced following an initialization run of 10 calls. In addition to the extra routines that have been described in the last few sections, there are some other changes that should be noted. The initialization routine sets to zero several new counters and also sets the header and trailer of the list to indicate that the list is initially empty. It will be recalled that the routine for creating a new call first checks that there are sufficient empty lines. This routine must now allow for the fact that some lines are occupied by waiting calls. The scan routine, finding the next call to finish, has also been slightly changed to allow for the presence of negative numbers in KEND.

When the program finds a line is busy, the call it is attempting to connect could have been removed from the waiting list. If so, service should be offered to the next call on the list, if there is one. The program goes to statement 71 when it finds a line is busy, to update 2 counters. Originally, the program then went to statement 10 to start the next event. It now goes to statement 79 to remove the first call from the list or return to 10 if the list is empty.

Finally, the routine for computing the output statistics and printing the output has been expanded. It now completes the computation needed to determine the link occupancy and includes a routine for printing out a table.

```
NO. OF LINES              50
NO. OF LINKS              10
MEAN CALL INTERVAL        12
MEAN CALL LENGTH         120
NUMBER OF CALLS         1000


RUN RESET AFTER    10  CALLS FINISHED


CLOCK TIME             10420

CALLS COMPLETED          672
BUSY CALLS               328
BLOCKED CALLS            307
LINK OCCUPANCY          7.87


    NO. IN TABLE         303

    MEAN VALUE          37.46

    STANDARD DEV.       30.72

    UPPER    OBSERVED     PER CENT   CUMULATIVE
    LIMIT    FREQUENCY    OF TOTAL   PERCENTAGE


         5       34         11.22       11.22
        10       29          9.57       20.79
        15       27          8.91       29.70
        20       17          5.61       35.31
        25       25          8.25       43.56
        30       23          7.59       51.16
        35       19          6.27       57.43
        40       19          6.27       63.70
        45       17          5.61       69.31
        50       15          4.95       74.26
        55        8          2.64       76.90
        60        9          2.97       79.87
        65        6          1.98       81.85
        70        9          2.97       84.82
        75        6          1.98       86.80
        80        6          1.98       88.78
        85        8          2.64       91.42
        90        2          0.66       92.08
        95        1          0.33       92.41
   OVERFLOW      23          7.59      100.00
```

Figure 10-11. Output from telephone system simulation.

Sec. 10-7 / Extension of the Telephone System Example

177

```
C      SIMULATION OF TELEPHONE SYSTEM - MODEL 2
C
C      PROGRAM REQUIRES INPUT CARDS WITH THE FOLLOWING INFORMATION
C      FIELD NO.    COLUMNS    INFORMATION
C
C      RUN CARD
C          1         1 -  6    CARD TYPE = 1
C          2         7 - 12    NUMBER OF TELEPHONE LINES
C          3        13 - 18    NUMBER OF LINKS
C          4        19 - 24    MEAN CALL INTER-ARRIVAL TIME
C          5        25 - 30    MEAN CALL LENGTH
C          6        31 - 36    NUMBER OF CALLS TO BE RUN
C          7        37 - 42    PRINT CONTROL  0 = PRINT
C                                             1 = NO PRINT
C      RESET CARD
C          1         1 -  6    CARD TYPE = 3
C          6        31 - 36    NUMBER OF CALLS TO BE RUN
C      TABLE CARD
C          1         1 -  6    CARD TYPE = 2
C          2         7 - 12    UPPER LIMIT OF 1ST INTERVAL
C          3        13 - 18    INTERVAL WIDTH
C          4        19 - 24    NO. OF INTERVALS
C      END CARD
C          1         1 -  6    CARD TYPE = 4
C
C      LOCATIONS IN THE PROGRAM HAVE THE FOLLOWING MEANINGS
C             IFROM  NEXT FUTURE CALL ORIGIN
C             ILNGTH NEXT FUTURE CALL LENGTH (IF PLACED)
C             ITA    NEXT FUTURE CALL ARRIVAL TIME
C             ITO    NEXT FUTURE CALL DESTINATION
C             KTFIN  TIME NEXT CALL IN PROGRESS FINISHES
C             KHEAD  POINTER TO FIRST WAITING CALL
C             KTAIL  POINTER TO LAST WAITING CALL
C             NARR   MEAN INTER-ARRIVAL TIME
C             NBLKS  NO OF BLOCKED CALLS
C             NBUSY  NO OF BUSY CALLS
C             NCOMP  NO OF COMPLETED CALLS
C             NERR   ERROR STOP LOCATION NUMBER
C             NFINS  NO OF CALLS PROCESSED
C             NLAST  LAST TIME NO. OF LINKS CHANGED
C             NLINES NO OF LINES
C             NLNGTH MEAN CALL LENGTH
C             NRUN   NO OF CALLS TO BE RUN
C             NSUM   ACCUMULATED LINKS OCCUPANCY COUNT
C             NWAIT  NUMBER OF BLOCKED CALLS WAITING
C             NUSE   NUMBER OF LINKS IN USE
C
C      GENERATE A RANDOM NUMBER - YFL
C
0001          SUBROUTINE RAND (IX,IY,YFL)
0002          IY = IX*1220703125
0003          IF (IY) 1,2,2
0004    1     IY = IY + 2147483647 + 1
0005    2     YFL = IY
0006          YFL = YFL*.4656613E-9
0007          IX = IY
0008          RETURN
0009          END

C      DEFINITION OF THE SYSTEM IMAGE
C
C             TABLE CALLED LINES HAS ONE WORD PER
C             LINE. VALUE IS 0/1 IF LINE IS FREE/BUSY
C             MINIMUM SIZE IS 2, MAXIMUM SIZE IS 100
C             TABLES CALLED KTO, KEND, KLNGTH AND KPNTR ARE FOR
C             CALLS IN PROGRESS. EACH ACTIVE CALL
C             IS PLACED AT THE LOCATION INDICATED
C             BY THE ORIGINATING LINE.
C             KTO    IS FOR THE CALLED PARTY
C             KEND IS TIME CALL WILL FINISH WHEN CALL IS CONNECTED AND
C             IS -(TIME OF ARRIVAL) WHEN CALL IS WAITING FOR CONNECTION
C             TABLE CALLED KLNGTH HOLDS LENGTH OF BLOCKED CALLS
C             TABLE CALLED KPNTR HOLDS POINTERS TO BLOCKED CALLS
C
```

Figure 10-10. Complete listing of extended telephone system simulation.

```
0001                DIMENSION LINES(100),KTO(100),KEND(100),ICARD(7),ITAB(100,2)
0002                DIMENSION KPNTR(100),KLNGTH(100)
             C
             C              INF IS THE LARGEST VALUE A VARIABLE CAN TAKE.
             C              KTFIN IS SET TO INF TO INDICATE AN EMPTY SYSTEM
             C
0003                NINTS = C
             C
             C       READ AND CHECK A CARD
             C
0004     60         READ (5,1,ERR = 107) (ICARD(I),I = 1,7)
0005      1         FORMAT (7I6)
0006                ITEM = ICARD(1)
0007                GO TO (61,65,611,82), ITEM
0008                GO TO 107
             C
             C              CHECK RESET CARD
             C
0009     611        IF (ICARD(6)) 107,107,612
0010     612        IF (NLINES) 105,105,613
0011     613        DO 615 I = 1,NLINES
0012                IF (KEND(I)) 616,615,614
0013     616        KEND(I) = KEND(I) + ICLOCK
0014                GO TO 615
0015     614        KEND(I) = KEND(I) - ICLOCK
0016     615        CONTINUE
0017                ITA = ITA - ICLOCK
0018                IF (INF - KTFIN) 618,618,617
0019     617        KTFIN = KTFIN - ICLOCK
0020     618        NSTR = NRUN
0021                NRUN = ICARD(6)
0022                NPRNT = 0
0023                GO TO 68
             C
             C              CHECK TABLE CARD
             C
0024     65         INTL = ICARD(3)
0025                LLIM = ICARD(2)
0026                NINTS = ICARD(4)
0027                NINTT = NINTS - 1
0028                IF (INTL) 107,60,63
0029     63         IF (LLIM) 107,69,69
0030     69         IF (NINTS - 3) 107,600,600
0031     600        IF (NINTS - 100) 601,601,107
0032     601        NTEMP = 0
0033                DO 602 I = 1,NINTT
0034                ITAB(I,1) = LLIM + NTEMP
0035     602        NTEMP = NTEMP + INTL
0036                GO TO 60
             C
             C              CHECK RUN CARD
             C
0037     61         DO 64 I = 2,6
0038                IF (ICARD(I)) 107,107,64
0039     64         CONTINUE
0040                NLINES = ICARD(2)
0041                NLNKS = ICARD(3)
0042                NARR = ICARD(4)
0043                NLNGTH = ICARD(5)
0044                NRUN = ICARD(6)
0045                NPRNT = ICARD(7)
0046                IF (NLINES - 2) 107,107,62
0047     62         IF (100 - NLINES) 107,67,67
             C
             C              CONVERT TO FLOATING POINT
             C
0048     67         ALINES = NLINES
0049                AARR = NARR
0050                ALNGTH = NLNGTH
             C
```

Figure 10-10. Complete listing of extended telephone system simulation (*cont.*).

```
        C       INITIALIZE PROGRAM
        C
0051            DO 66 I = 1,NLINES
CC52            LINES(I) = C
JJ53            KTO(I) = 0
CC54            KLNGTH(I) = 0
0055            KPNTR(I) = 0
CC56      66    KENC(I) = 0
CC57            NSTR = C
CJ5E            NWAIT = 0
JJ59            NLSE = C
OCEC            NFIN = 0
CC61            INF = 2147483647
CC62            KHEAD = INF
JJE3            KTAIL = INF
JJ64            KTFIN = INF
CC65            IX = 13579
        C
        C             ENTER HERE FOR RESET
        C
CC66      6F    ICLOCK = 0
0067            NBLKS = C
CC68            NBUSY = 0
CC65            NCOMP = 0
007C            NFINS = C
OC71            NLAST = 0
0072            NSUM = 0
JJ73            IF (NINTS) 107,606,603
CO74     603    DC 604 I = 1,NINTS
CC75     604    ITAB(I,2) = 0
CC76            TTCL = C.0
0C77            TSUM = 0.0
CJ78            TSSQ = 0.0
        C
        C             COMPUTE FIRST ARRIVAL TIME
        C
CC79     606    IF (ICARD(1) - 3) 605,10,605
0C80     605    CALL RAND (IX,IY,YFL)
JJ81            ITA = -AARR*ALOG(YFL) + ICLOCK
JJ82            GC TO 10
        C
        C       FIND NEXT POTENTIAL EVENT
        C
0083      1C    IF (KTFIN - INF) 11,13,103
0084      11    IF (KTFIN - ITA) 14,14,13
CCE5      13    ICLCCK = ITA
JJ86            IF (ICLCCK) 1C2,18,18
CC87      14    ICLCCK = KTFIN
CC88            GO TO 50
        C
        C             CREATE A NEW CALL
        C
0089      18    IF (NLINES - 2*NUSE - NWAIT - 2) 605,6C5,15
0C90      15    CALL RAND (IX,IY,YFL)
0091            IFROM = ALINES*YFL + 1
0092            IF (LINES(IFROM)) 104,16,15
0093      16    CALL RAND (IX,IY,YFL)
CC94            ITC = ALINES*YFL + 1
JJ95            IF (IFRCM - ITO) 17,16,17
CLS6      17    CALL RAND (IX,IY,YFL)
JJ97            ILNGTH = -ALNGTH*ALOG(YFL)
CO98            LINES(IFROM) = 1
        C
        C             CCMPUTE NEXT ARRIVAL TIME
        C
CCS9            CALL RAND (IX,IY,YFL)
01CC            ITA = -AARR*ALOG(YFL) + ICLOCK
01J1            GO TO 30
        C
        C       DISCONNECT A CALL
        C
```

Figure 10-10. Complete listing of extended telephone system simulation (*cont.*).

```
                  C                    NF IN HAS LOCATION OF CALL
0102            50          LINES(NFIN) = 0
0103                        NTEMP = KTO(NFIN)
01C4                        LINES(NTEMP) = 0
0105                        NUSE = NUSE - 1
0106                        KTC(NFIN) = 0
01C7                        KEND(NFIN) = 0
                  C
                  C                    UPDATE NEXT-TO-FINISH RECORDS
                  C
0108            51          KTFIN = INF
0109                        IF (NUSE) 101,73,52
011C            52          DO 55 I = 1,NLINES
0111                        IF (KEND(I)) 55,55,53
0112            53          IF (KEND(I) - KTFIN) 54,55,55
0113            54          KTFIN = KEND(I)
0114                        NFIN = I
0115            55          CONTINUE
0116                        GO TO 73
                  C
                  C          TEST IF CALL CAN BE MADE
                  C
0117            30          IF (NLNKS - NUSE) 101,70,31
0118            31          IF (LINES(ITO)) 104,20,71
                  C
                  C          CONNECT A CALL
                  C
0119            20          LINES(ITO) = 1
0120                        NUSE = NUSE + 1
0121                        KTC(IFROM) = ITO
0122                        KEND(IFROM) = ICLOCK + ILNGTH
0123                        IF (KEND(IFROM) - KTFIN) 21,72,72
0124            21          KTFIN = KEND(IFROM)
0125                        NFIN = IFROM
0126                        GO TO 72
                  C
                  C          ADD CALL TO END OF BLOCKED CALL CHAIN
                  C
0127            70          NBLKS = NBLKS + 1
0128                        NWAIT = NWAIT + 1
0129                        IF (INF - KTAIL) 110,702,701
0130            701         KPNTR(KTAIL) = IFROM
0131                        GO TO 703
0132            702         KHEAD = IFROM
0133            703         KPNTR(IFROM) = INF
0134                        KTAIL = IFROM
0135                        KTO(IFROM) = ITO
0136                        KLNGTH(IFROM) = ILNGTH
0137                        KEND(IFROM) = -ICLOCK
0138                        GO TO 10
                  C
                  C          GATHER STATISTICS
                  C
0139            71          NBUSY = NBUSY + 1
0140            74          NFINS = NFINS + 1
0141                        LINES(IFROM) = 0
0142                        IF (NRUN - NFINS) 80,80,79
0143            72          NSUM = NSUM + (ICLOCK - NLAST)*(NUSE - 1)
0144                        NLAST = ICLOCK
0145                        GO TO 10
0146            73          NSUM = NSUM + (ICLOCK - NLAST)*(NUSE + 1)
0147                        NLAST = ICLOCK
0148                        NCOMP = NCOMP + 1
0149                        NFINS = NFINS + 1
0150                        IF (NRUN - NFINS) 80,80,79
                  C
```

Figure 10-10. Complete listing of extended telephone system simulation (*cont.*).

```
                    C        REMOVE FIRST CALL FROM BLOCKED CALL CHAIN
                    C
0151        79               IF (INF - KHEAD) 110,10,704
0152        704              IFROM = KHEAD
0153                         ITO = KTO(IFROM)
0154                         ILNGTH = KLNGTH(IFROM)
0155                         NWAIT = NWAIT - 1
0156                         INPT = ICLOCK + KEND(IFROM)
0157                         IF (KHEAD - KTAIL) 705,706,705
0158        705              KHEAD = KPNTR(IFROM)
0159                         GO TO 707
0160        706              KHEAD = INF
0161                         KTAIL = INF
0162        707              KTO(IFROM) = 0
0163                         KEND(IFROM) = 0
0164                         KLNGTH(IFROM) = 0
0165                         KPNTR(IFROM) = 0
                    C
                    C        TABULATE WAITING TIME
                    C
0166                         IF (NINTS) 107,30,708
0167        708              DO 77 I = 1,NINTT
0168                         IF (INPT - ITAB(I,1)) 76,76,77
0169        76               ITAB(I,2) = ITAB(I,2) + 1
0170                         GO TO 78
0171        77               CONTINUE
0172                         ITAB(NINTS,2) = ITAB(NINTS,2) + 1
0173        78               TTOL = TTOL + 1.0
0174                         TSUM = TSUM + INPT
0175                         TSSQ = TSSQ + INPT**2
0176                         GO TO 30
                    C
                    C        COMPUTE FINAL STATISTICS AND PRINT OUT
                    C
0177        80               IF (NPRNT) 60,801,60
0178        801              NSUM = NSUM + (ICLOCK - NLAST)*NUSE
0179                         ASUM = NSUM
0180                         ACLOCK = ICLOCK
0181                         IF (ACLOCK) 108,108,81
0182        81               AOCCY = ASUM/ACLOCK
0183                         WRITE(6,7)NLINES,NLNKS,NARR,NLNGTH,NRUN
0184        7                FORMAT(1H1,9X,12HNO. OF LINES,12X,I6/10X,12HNO. OF LINKS,
                            112X,I6/10X,18HMEAN CALL INTERVAL,7X,I5/
                            210X,16HMEAN CALL LENGTH,9X,I5/10X,15HNUMBER OF CALLS,9X,I6///)
0185                         IF (NSTR) 105,803,802
0186        802              WRITE(6,95)NSTR
0187        95               FORMAT(10X,15HRUN RESET AFTER,I5,16H  CALLS FINISHED//)
0188        803              WRITE(6,2)ICLOCK,NCOMP,NBUSY,NBLKS,AOCCY
0189        2                FORMAT(10X,10HCLOCK TIME,11X,I9//10X,15HCALLS COMPLETED,
                            16X,I9/10X,10HBUSY CALLS,11X,I9/10X,13HBLOCKED CALLS,8X,
                            2I9/10X,14HLINK OCCUPANCY,6X,F10.2)
                    C
                    C        COMPUTE AND WRITE OUT TABLE
                    C
0190                         IF (NINTS) 109,60,84
0191        84               ITOL = TTOL
0192                         IF (TTOL) 85,85,86
0193        85               TMEAN = 0.0
0194                         TSTD = 0.0
0195                         GO TO 89
0196        86               IF (TTOL - 1.0) 87,87,88
0197        87               TMEAN = TSUM
0198                         TSTD = 0.0
0199                         GO TO 89
0200        88               TMEAN = TSUM/TTOL
0201                         TSTD = ((TSSQ - TSUM**2/TTOL)/(TTOL - 1.0))**0.5
0202        89               WRITE(6,6)ITOL,TMEAN,TSTD
0203        6                FORMAT(///15X,12HNO. IN TABLE,3X,I10//15X,10HMEAN VALUE,5X,
                            1F10.2//15X,13HSTANDARD DEV.,2X,F10.2//15X,17HUPPER    OBSERVED,
                            224H     PER CENT  CUMULATIVE/15X,17HLIMIT    FREQUENCY,
                            324H    OF TOTAL   PERCENTAGE//)
0204                         ATEMP1 = 0.0
0205                         DO 83    I = 1,NINTT
0206                         NTEMP = ITAB(I,2)
```

Figure 10-10. Complete listing of extended telephone system simulation (*cont.*).

```
0207            ATEMP = NTEMP
0208            IF (TTOL) 83,83,806
0209      806   ATEMP = ATEMP*100.0/TTOL
0210            ATEMP1 = ATEMP1 + ATEMP
0211      83    WRITE(6,3)ITAB(I,1),ITAB(I,2),ATEMP,ATEMP1
0212      3     FORMAT(11X,I9,3X,I9,3X,F9.2,3X,F9.2)
0213            NTEMP = ITAB(NINTS,2)
0214            ATEMP = NTEMP
0215            IF (TTOL) 804,804,805
0216      805   ATEMP = ATEMP*100.0/TTOL
0217            ATEMP1 = ATEMP1 + ATEMP
0218      804   WRITE(6,9)NTEMP,ATEMP,ATEMP1
0219      9     FORMAT(12X,8HOVERFLOW,3X,I9,3X,F9.2,3X,F9.2)
0220            GO TO 60
          C
0221      82    CALL EXIT
          C
          C     ERROR ROUTINE
          C
0222      101   NERR = 1
0223            GO TO 90
0224      102   NERR = 2
0225            GO TO 90
0226      103   NERR = 3
0227            GO TO 90
0228      104   NERR = 4
0229            GO TO 90
0230      108   NERR = 8
0231            GO TO 94
0232      109   NERR = 9
0233            GO TO 94
0234      110   NERR = 10
0235      90    WRITE(6,8)NERR
0236      8     FORMAT(//10X,13HERROR STOP NO,5X,I5//)
0237      91    WRITE(6,4)NUSE,KTFIN,IFROM,ITO,ITA,KHEAD,KTAIL
0238      4     FORMAT(7I6//)
0239            DO 92 I = 1,NLINES
0240            WRITE(6,5)I,LINES(I),KTO(I),KEND(I),KLNGTH(I),KPNTR(I)
0241      5     FORMAT(5X,6(5X,I6))
0242      92    CONTINUE
0243            GO TO 80
0244      105   WRITE(6,96)
0245      96    FORMAT(//10X,18HILLEGAL RESET CARD//)
0246            GO TO 60
0247      107   WRITE(6,93)
0248      93    FORMAT(//10X,16HINPUT CARD ERROR//)
0249      920   FORMAT(10X,I6)
0250            GO TO 60
0251      94    WRITE(6,8)NERR
0252            GO TO 60
0253            END
```

Figure 10-10. Complete listing of extended telephone system simulation (*cont.*).

10-8
Simultaneous Events

In simulation programs, attention must be paid to conditions that exist when two or more events are scheduled to occur simultaneously. While this means that the events occur at the same instant in the system, the simulation program is forced to execute the events in sequence. There may be priorities assigned as attributes of the entities; in which case, the scan of events considers events in order of priority.

If no particular rule of ordering is made, the sequence of execution is, in effect, unpredictable. The choice of order may not be significant in the over-all representation of the system. However, in trying to follow the detailed performance of the model, particularly when debugging the model, it may be difficult to interpret the simulated action. A study of the event history, with a detailed understanding of how the simulation algorithm operates, may enable the order to be determined. However, as a practical matter, the precise sequence of events may be so involved that it is virtually impossible to untangle. For this reason, priority is sometimes used to control the ordering of simultaneous events, even though there may not be any corresponding distinction in the system.

The events that occur simultaneously may not be independent of each other. For example, two entities may be competing for the same item of equipment. The system can behave very differently according to the choice of order. Two simulation runs which logically appear to be exactly the same can give very different results. Some slight difference in the way in which the programs are assembled or executed can cause events to occur in a different order. In particular, when random numbers are being used, differences in random number sequences may make significant differences between otherwise identical runs.

Another aspect of simultaneous events that must be considered is the fact that an entity moving through the system may be involved in activities that require no time. There can, in fact, be a string of such "zero-time" activities. The execution of an event involving a zero-time activity immediately produces another event due for execution at the same clock time. By itself, this may present no difficulty; but, if there are several different entities, all due to participate in events simultaneously, and any of them encounters zero-time activities, the question of choosing the order of the scanning can be important.

The condition is illustrated in Fig. 10-12, which represents a set of entities $E_1, E_2, \ldots, E_n$ all due to begin moving at the *same* clock time and each headed for individual strings of zero-time activities of various lengths. The activities to be executed during this clock instant for the ith entity are A_{ij} ($j = 1, 2, \ldots, k_i$), so that potentially there is a total of $\sum_{i=1}^{n} k_i$ events due for execution at the same instant. If no control is exercised over the order in which events are scanned, the order of execution will be unpredictable, leading to possible confusion and violation of priority rules.

Two simple alternative scanning rules may be invoked to establish a predictable control. Each entity may be taken through all potential activities that can occur at the same clock instant, before turning attention to another entity. Alternatively, each entity may be allowed one activity at a time before the scan moves to the next entity. In both cases, the order of movement between entities is controlled (if necessary) by priority rules. Referring to Fig. 10-12, it is seen that if the events are considered to be ranked in order of priority from

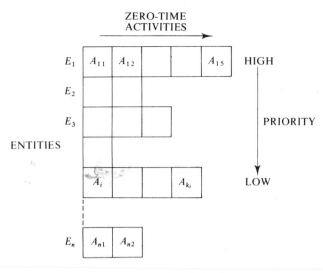

Figure 10-12. Representation of simultaneous event.

top to bottom so that scanning is from top to bottom, the first scan rule is equivalent to scanning the matrix of events by row from top to bottom. The second rule is equivalent to scanning by column from left to right. The first rule of scanning by row is generally preferred because it is easier to interpret the sequence of events executed by the program. It is not always possible to determine how many entities are due to move simultaneously, so that the sequence of events in a column scan is difficult to follow.

10-9
Blocked Events

When an event cannot be executed at its scheduled time, it is said to be blocked, and the record for the event must lie dormant until the condition causing the blocking changes. The records of the blocked events are often kept in a separate list called a *ready list*, and the simplest programming arrangement is to rescan the ready list whenever the state of the system changes, to determine whether conditions have become favorable for any of the events. There can be a chain effect, with the execution of one event releasing one or more events at the same clock time, which in turn may release others, and so on, leading again to the problem of considering the order of processing simultaneous events.

To describe the problem involved, consider the situation shown in Fig. 10-13, which represents a ready list of blocked event records. The records are arranged in an order of priority and are scanned from the highest to the lowest priority (top to bottom). When the scan of the list is started, it may read a record, say C, which is found to represent an event that has become released. This release may, in turn, allow the events represented by records B and F to be released. Record B has been passed by the scan and as a result, if the scan continues down the ready list, the lower priority record, F, may take advantage of the change ahead of B. For complete accuracy, the ready list scan should be restarted whenever any change in the list occurs, and the process repeated with each detected change until the scan manages to get through the entire list without any further change.

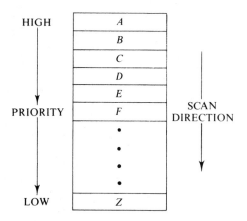

Figure 10-13. Scanning blocked events.

The situation is further complicated by the fact that the updating of clock time introduces new untried events for examination, and the priority of these new events relative to events in the ready list also needs to be considered. A general solution to the problem is illustrated in Fig. 10-14. When all activity at one clock time is finished, the scan routine updates the clock time to the time of the next future event record. The record is then merged, according to its priority, with the ready list, which contains records of events that remained blocked from the last scan. If there should be more than one future event record at the new clock time, they are all merged with the ready list before proceeding further.

A scan of the ready list is then started, beginning from the highest priority record. Whenever an entry is found that can be executed, the scan returns to the beginning of the list to restart the scan. When the scan finally scans the last entry without it moving, all events that are possible at the current time have been performed and the scanning program can move to update the clock time again. If the rule for handling zero-time activities is to process all zero-time activities for one entity before proceeding to the next entity, a check is made for zero-time activity before returning to the scan, as illustrated in Fig. 10-14. If the rule is to allow only one move for an entity, the return to the scan is made immediately after the move.

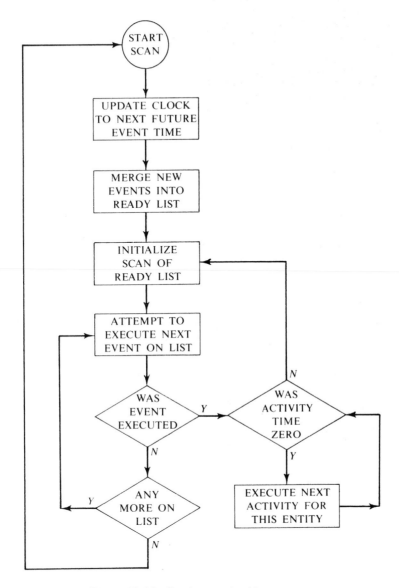

Figure 10-14. Simple scan algorithm.

The repeated retrial of events involved in the scanning can be very time consuming. The process can be made more efficient if it is possible to identify the particular activity that will result in releasing a blocked event. Blocked events can then be made inactive, either by removing them from the ready list, or by setting a status flag in the event record and arranging that the event is not retried until the flag is reset. The activity routine that will release the event can then be programmed to return the event to the ready list or reset the status flag. A better efficiency is obtained if the activity routine also sets a change flag indicating that an event in the list has been reactivated. When the scan has finished processing an event, it can then check the change flag and re-initiate the scan only if it has been set. Figure 10-15 shows a flow chart of this algorithm.

The GPSS program discussed in the next chapter uses a scanning algorithm that is essentially the same as that given in Fig. 10-15. Two main lists of event records (transactions) are kept; one for future events maintained in chronological order and the other a ready list (called in GPSS, the current events chain) of events that are currently blocked. While remaining on the ready list, event records are also put onto one of many sublists, joining all events awaiting a particular change of conditions. When that change occurs, the program uses the sublist to reset the status flags of the unblocked events. It also sets the change flag on, so that the scanning algorithm will immediately retry the events. Each activity routine (block type routine) is arranged to set the status flag of any event it is unable to execute.

Exercises

Explain the changes you would make and write the programming steps to modify the telephone system simulation in the following ways:

10-1 Tabulate the waiting time only for calls that eventually get connected.

10-2 Keep records of the individual links being used.

10-3 Derive the average number of calls made to a line.

10-4 Make the selection from the waiting calls random.

10-5 Keep the waiting calls ordered in time of arrival but offer service first to any call going to line number 1, if line number 1 is free. If there is no call to line 1 waiting or line 1 is busy, offer service to the call that has waited longest.

10-6 Give service to waiting calls in the order of the line number that the call comes from.

10-7 Make the list of waiting calls a double-threaded list.

Bibliography

1 Neely, Peter M., "Comparison of Several Algorithms for Computation of Means, Standard Deviations and Correlation Coefficients," *Comm. ACM*, IX, No. 7 (July, 1966), 496–499.

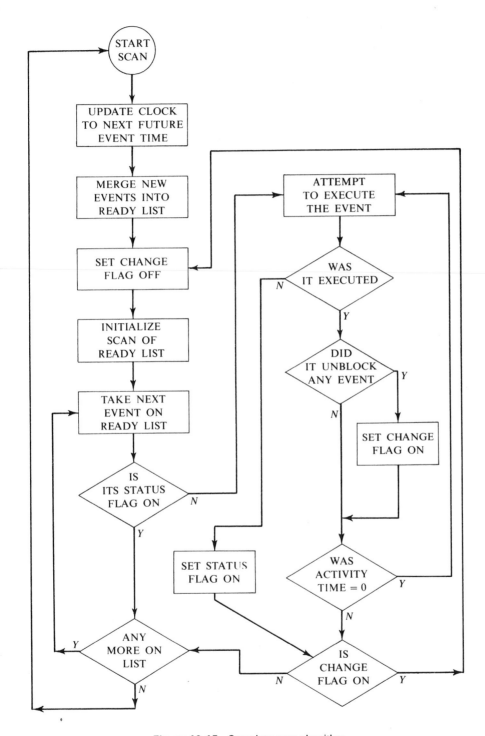

Figure 10-15. Complete scan algorithm.

2 Perlis, A. J., and Charles Thornton, "Symbol Manipulation by Threaded Lists," *Comm. ACM*, III, No. 4 (April, 1960), 195–204.

3 Madnick, Stuart E., "String Processing Techniques," *Comm. ACM*, X, No. 7 (July, 1967), 420–424.

11

INTRODUCTION
TO GPSS

The last two chapters have discussed the use of a general purpose scientific programming language for simulating discrete systems. In this and the next three chapters, we discuss programming languages designed specifically for this purpose. The first program to be discussed is called the General Purpose Simulation System, or GPSS. Several versions have been written (1), (2), (3). The one that will be described is GPSS/360; see (4), (5). For brevity it will be referred to as GPSS.

The system to be simulated in GPSS is described as a block diagram in which the blocks represent the activities, and lines joining the blocks indicate the sequence in which the activities can be executed. Where there is a choice of activities, more than one line leaves a block and the condition for the choice is stated at the block.

The use of block diagrams to describe systems is, of course, very familiar. However, the form taken by a block diagram description usually depends upon the person drawing the block diagram. To base a programming language on this descriptive method, each block must be given a precise meaning. The approach taken in GPSS is to define a set of 37 specific *block types*, each of which represents a characteristic action of systems. The program user must draw a block diagram of the system using only these block types.

Each block type is given a name that is descriptive of the block action and is represented by a particular symbol. Figure 11-1 shows the symbols used for

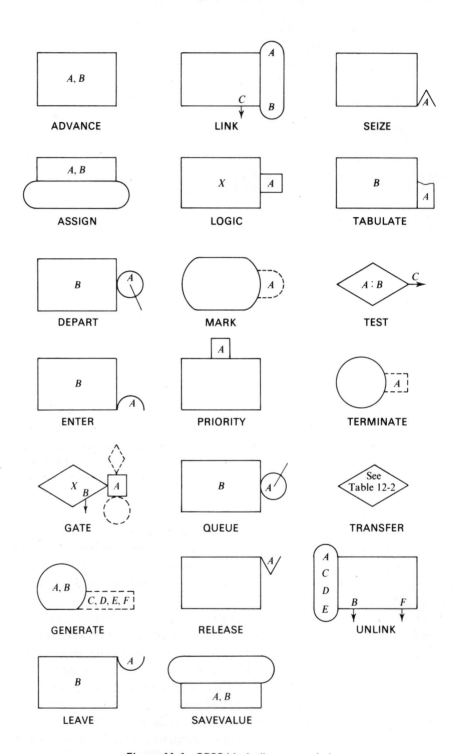

Figure 11-1. GPSS block-diagram symbols.

Table 11-1 GPSS Block Types

Operation	A	B	C	D	E	F
ADVANCE	Mean	Modifier				
ASSIGN	Param. No. (±)	Source				
DEPART	Queue No.	(Units)				
ENTER	Storage No.	(Units)				
GATE†	Item No.	(Next block B)				
GENERATE	Mean	Modifier	(Offset)	(Count)	(Priority)	(Params.)
LEAVE	Storage No.	(Units)				
LINK	Chain No.	Order	(Next block B)			
LOGIC {R S I}	Switch No.					
MARK	(Param. No.)					
PRIORITY	Priority					
QUEUE	Queue No.	(Units)				
RELEASE	Facility No.					
SAVEVALUE	S.V. No. (±)	SNA				
SEIZE	Facility No.					
TABULATE	Table No.	(Units)				
TERMINATE	(Units)					
TEST†	Arg. 1	Arg. 2	(Next block B)			
TRANSFER	Select. Factor	Next block A	Next block B			
UNLINK	Chain No.	Next block A	Count	(Param. No.)	(Arg.)	(Next block B)

† Columns 13 and 14. See Sec. 12-7 for codes.
( ) indicates optional field.

the block types that will be described in this and the next chapter. To assist the reader, the block diagrams drawn in this chapter will name the block types, although this is not usually done by a programmer familiar with GPSS. Coding instructions for all the described block types are similarly brought together in Table 11-1. Table 11-2 describes the control cards. Each block type has a number of data fields. As the blocks are described, the fields will be referred to as field A, B, C, and so on, reflecting the order in which they are key-punched.

Table 11-2 GPSS Control Cards

Location	Operation	A	B	C	D
	CLEAR				
	END				
Function No.	FUNCTION	Argument	$\begin{Bmatrix} C \\ D \\ L \end{Bmatrix}$ No. of Points		
	INITIAL	Xn	Value		
	JOB				
	RESET				
	SIMULATE				
	START	Run Count	(NP)		
Storage No.	STORAGE	Capacity			
Table No.	TABLE	Argument	Lower limit	Interval	No. of intervals

Moving through the system being simulated are entities that depend upon the nature of the system. For example, a communication system is concerned with the movement of messages, a road transportation system with motor vehicles, and a data processing system with records, and so on. In the simulation, these entities are called *transactions*. The sequence of events in real time is reflected in the movement of transactions from block to block in simulated time.

Transactions are created at one or more **GENERATE** blocks and are removed from the simulation at **TERMINATE** blocks. There can be many transactions simultaneously moving through the block diagram. Each transaction is always positioned at a block and most blocks can hold many transactions simultaneously. The transfer of a transaction from one block to another occurs instantaneously at a specific time or when some change of system condition occurs.

A GPSS block diagram can consist of many blocks up to some limit prescribed by the program (usually set to 1,000). An identification number called a *location* is given to each block, and the movement of transactions is usually from one block to the block with the next highest location. The locations are assigned automatically by an assembly program within GPSS so, when a problem is coded, the blocks are listed in sequential order. Blocks that need to be identified in the programming of problems (for example, as points to which a transfer is to be made) are given a symbolic name. The assembly program will associate the name with the appropriate location. Symbolic names of blocks and other entities of the program must be from three to five non-blank characters of which *the first three must be letters.*

11-2

Action Times

Clock time is represented by an integral number, with the interval of real time corresponding to a unit of time chosen by the program user. The unit of time is not specifically stated but is implied by giving all times in terms of the same unit. One block type called ADVANCE is concerned with representing the expenditure of time. The program computes an interval of time called an *action time* for each transaction as it enters an ADVANCE block, and the transaction remains at the block for this interval of simulated time before attempting to proceed. The only other block type that employs action time is the GENERATE block, which creates transactions. The action time at the GENERATE block controls the *interval* between successive arrivals of transactions.

The action time may be a fixed interval (including zero) or a random variable, and it can be made to depend upon conditions in the system in various ways. An action time is defined by giving a *mean* and *modifier* for the block. If the modifier is zero, the action time is a constant equal to the mean. If the modifier is a positive number ($\leq$ mean), the action time is an integer random variable chosen from the range mean $\pm$ modifier, with equal probabilities given to each number in the range. Sometimes, this uniform distribution is an accurate representation of a random process in the system, but the principal purpose in providing this way of representing a random time is to allow for cases where randomness is known to exist but no detailed information is available on the probability distribution.

It is possible to introduce a number of *functions* which are tables of numbers relating an input variable to an output variable. Details of the functions are given in Sec. 12-3. By specifying the modifier at an ADVANCE or GENERATE block to be a function, the value of the function controls the action time. The action time is derived by *multiplying* the mean by the value of the function. Various types of input can be used for the functions, allowing the functions to introduce a variety of relationships between the variables of a system. In particular, by making the function an inverse cumulative probability distribution, and using as input a random number which is uniformly distributed, the function can provide a stochastic variable with a particular non-uniform distribution in the manner described in Sec. 6-9.

11-3
Succession of Events

The program maintains records of when each transaction in the system is due to move. It proceeds by completing all movements that are scheduled for execution at a particular instant of time and can logically be performed. Where there is more than one transaction due to move, the program processes transactions in the order of their priority class, and on a first-come, first-served basis within priority class.

Normally, a transaction spends no time at a block other than at an ADVANCE block. Once the program has begun moving a transaction, therefore, it continues to move the transaction through the block diagram until one of several circumstances arises. The transaction may enter an ADVANCE block with a non-zero action time, in which case, the program will turn its attention to other transactions in the system and return to that transaction when the action time has been expended. Secondly, the conditions in the system may be such that the action the transaction is attempting to execute by entering a block cannot be performed at the current time. The transaction is said to be blocked and it remains at the block it last entered. The program will automatically detect when the blocking condition has been removed and will start to move the transaction again at that time. A third possibility is that the transaction enters a TERMINATE block, in which case it is removed from the simulation. A fourth possibility is that a transaction may be put on a chain. This concept will be explained in the next chapter.

When the program has moved one transaction as far as it can go, it turns its attention to any other transactions due to move at the same time instant. If all such movements are complete, the program advances the clock to the time of the next most imminent event and repeats the process of executing events.

11-4
Choice of Paths

The TRANSFER block allows some location other than the next sequential location to be selected. The choice is normally between two blocks referred to as next blocks *A* and *B* (the terms exits 1 and 2 are also used). The method used for choosing is indicated by a *selection factor* in field A of the TRANSFER block. It can be set to indicate one of nine choices. Next blocks *A* and *B* are placed in fields B and C, respectively. If no choice is to be made, the selection factor is set to zero (or not punched). An unconditional transfer is then made to next block *A*.

Of the other modes of choice, only two will be described here. Others will be described in Sec. 12-4 (see also Table 12-2). A random choice can be made by setting the selection factor, S, to a three-digit decimal fraction. The probability of going to next block A is then $1 - S$, and to the next block B is S. A conditional mode, indicated by setting field A to BOTH, allows a transaction to select an alternate path depending upon existing conditions. The transaction goes to next block A if this move is possible, and to next block B if it is not. If both moves are impossible, the transaction waits for the first to become possible, giving preference to A in the event of simultaneity.

11-5

Simulation of a Manufacturing Shop

To illustrate the features of the program described so far, consider the following simple example. A machine tool in a manufacturing shop is turning out parts at the rate of one every 5 minutes. As they are finished, the parts go to an inspector, who takes 4 ± 3 minutes to examine each one and rejects about 10% of the parts. Each part will be represented by one transaction, and the time unit selected for the problem will be 1 minute.

A block diagram representing the system is shown in Fig. 11-2. The usual convention used in drawing blocks is to place the block location (where needed) at the top of the block; the action time is indicated in the center in the form $T = a, b$, where a is the mean and b is the modifier; and the selection factor is placed at the bottom of the block.

A GENERATE block is used to represent the output of the machine by creating one transaction every five units of time. An ADVANCE block with a mean of 4 and modifier of 3 is used to represent inspection. The time spent on

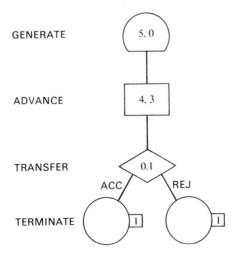

Figure 11-2. Manufacturing shop—model 1.

inspection will therefore be any one of the values 1, 2, 3, 4, 5, 6, or 7, with equal probability given to each value. Upon completion of the inspection, transactions go to a TRANSFER block with a selection factor of 0.1 so that 90% of the parts go to the next location (exit 1) called ACC, to represent accepted parts and 10% go to another location (exit 2) called REJ to represent rejects. Since there is no further interest in following the history of the parts in this simulation, both locations reached from the TRANSFER block are TERMINATE blocks.

The problem is coded on a special GPSS coding form as shown in Fig. 11-3. Column 1 is only used for a comment card. An * in column 1 results in the entire card being printed. A field from columns 2 to 6 contains the location of the block where it is necessary for it to be specified. The GPSS program will automatically assign sequential location numbers as it loads the card, so it is not usually necessary for the user to assign locations. The TRANSFER block, however, will need to make reference to the TERMINATE blocks to which it sends transactions, so these blocks have been given symbolic location names ACC and REJ.

The second field of the coding form, from columns 8 to 18, contains the block type name which must begin in column 8. Beginning at column 19, a

IBM GPSS III - CODING FORM

Printed in U.S.A. X20-1701-0

SHEET__OF__

NAME	DATE	PROBLEM NO.	PHONE NO.	PROGRAM TITLE

```
LOCATION  OPERATION   A,B,C,D,E,F ------>
*                     MANUFACTURING SHOP - MODEL 1
*
          GENERATE    5                   CREATE PARTS
          ADVANCE     4,3                 INSPECT
          TRANSFER    .1,ACC,REJ          SELECT REJECTS
 ACC      TERMINATE   1                   ACCEPTED PARTS
 REJ      TERMINATE   1                   REJECTED PARTS
          START       1000
```

Figure 11-3. Coding of model 1.

series of fields may be present, each separated by commas and having no inbedded blanks. Anything following the first blank is treated as a comment. The meaning of the fields depends upon the block type. Table 11-1 summarizes the information for all the block types that will be described. The table includes references to some block types that will be described in the next chapter.

In the case of both the GENERATE and the ADVANCE blocks, fields A and B hold respectively the mean and modifier controlling the action time. Field C of the GENERATE block is an offset time, determining the time at which the first transaction arrival time will be generated. If it is zero, the first transaction is generated from time equal to 1. The mean at a GENERATE block can be zero. If so, the user must arrange for the flow of transactions to be controlled by system conditions. Field D of a GENERATE block is a count that determines the total number of transactions the block will create. If it is zero, there is no limit. In the present example, the modifier at the GENERATE block is zero and no field higher than the A field is needed. Any fields undefined at the time the first blank is met are automatically set to zero.

For the TRANSFER block, the first field is the selection factor and the B and C fields are exits 1 and 2, respectively. In this case, exit 1 is the next sequential block, and it would be permissible to omit the name ACC in both the TRANSFER field B and the location field of the first TERMINATE block. Both commas must, however, be punched to show that field B is missing so that the TRANSFER block would then be coded: TRANSFER .1,,REJ. It should also be noted that when a TRANSFER block is used in an unconditional mode, so that field A is zero or not punched, a comma must still be included to indicate the field. Thus an unconditional transfer to REJ would be coded TRANSFER 0, REJ or TRANSFER ,REJ.

The program runs until a certain count is reached as a result of transactions terminating. Field A of the TERMINATE block carries a number indicating by how much the termination count should be incremented when a transaction terminates at that block. The number must be positive and it can be zero but there must be at least one TERMINATE block that has a non-zero field A. In this case, both TERMINATE blocks have 1, so the terminating counter will add up to the total number of transactions that terminate; in other words, the total number of both good and bad parts inspected.

The last line shown in Fig. 11-3 is for a control card called START, which indicates the end of the problem deck and contains in field A the value the terminating counter is to reach to cause the simulation run to end. In this case the START card is set to stop the simulation at 1,000.

One card is punched for each statement of the problem, and the deck of cards is made the input to the GPSS program deck. The program reads the cards, executes the simulation, and then prints an output report. For the problem just programmed, the complete output appears as shown in Fig. 11-4.

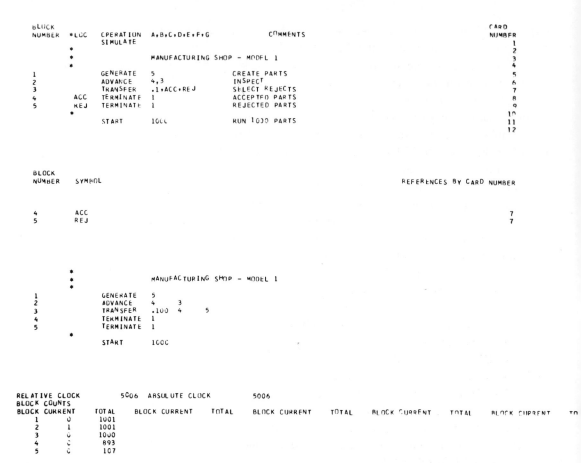

```
BLOCK                                                                                           CARD
NUMBER   *LOC   CPERATION  A,B,C,D,E,F,G              COMMENTS                                   NUMBER
                SIMULATE                                                                         1
         *                                                                                      2
         *                 MANUFACTURING SHOP - MODEL 1                                          3
         *                                                                                      4
1               GENERATE   5                          CREATE PARTS                              5
2               ADVANCE    4,3                         INSPECT                                   6
3               TRANSFER   .1,ACC,REJ                  SELECT REJECTS                            7
4        ACC    TERMINATE  1                           ACCEPTED PARTS                            8
5        REJ    TERMINATE  1                           REJECTED PARTS                            9
         *                                                                                      10
                START      1000                        RUN 1000 PARTS                            11
                                                                                                12


BLOCK
NUMBER   SYMBOL                                                         REFERENCES BY CARD NUMBER


4        ACC                                                                              7
5        REJ                                                                              7




         *
         *                 MANUFACTURING SHOP - MODEL 1
         *
1               GENERATE   5
2               ADVANCE    4        3
3               TRANSFER   .100     4        5
4               TERMINATE  1
5               TERMINATE  1
         *
                START      1000


RELATIVE CLOCK       5006   ABSOLUTE CLOCK         5006
BLOCK COUNTS
BLOCK CURRENT     TOTAL    BLOCK CURRENT    TOTAL    BLOCK CURRENT    TOTAL    BLOCK CURRENT    TOTAL    BLOCK CURRENT   TO
1       0         1001
2       1         1001
3       0         1000
4       0          893
5       0          107
```

Figure 11-4. Output for model 1.

The original input deck is first printed, with the locations assigned by the program listed to the left and a sequential card number on the right. A table of symbolic locations is then given showing the location assigned to each symbol. To help the user, the number of the cards that make reference to a symbol is also given. A listing of the assembled problem follows with all symbolic references replaced by their locations.

In the remaining examples, the full output will not be given. Only the initial listing will be shown. Note that the location numbers that appear on the left are placed there by the program; they are not punched in the input deck.

The first line of output following the listings gives the time at which the simulation stopped. The meaning of the absolute and relative times, which in this case have the same value, will be explained in Sec. 11-9. The time is followed by a listing of block counts. Two numbers are shown for each block of the model that was entered by a transaction. On the left is a count of how many transactions were in the block at the time the simulation stopped, and on the right is a figure showing the total number of transactions that entered the block during the simulation. The results show that the total counts at blocks 4 and 5 were 893 and 107 respectively, showing that of the 1,000 parts inspected, 89.3% were accepted and 10.7% were rejected. In the present example, this is the only output given. Other types of output will be described as more details of the program are explained.

Associated with the system being simulated are many permanent entities, such as items of equipment, which operate on the transactions. Two types of permanent entities are defined in GPSS to represent system equipment.

A *facility* is defined as an entity that can be engaged by a single transaction at a time. A *storage* is defined as an entity that can be occupied by many transactions at a time, up to some predetermined limit. There can be many instances of each type of entity to a limit set by the program (usually 300). Individual entities are identified by number; a separate number sequence being used for each type. The number 0 for these, and all other GPSS entities, is illegal. The user may assign the numbers, in any order, or he may use symbolic names and let the assembly program assign the numbers. Some examples of how the system entities might be interpreted in different systems are:

Type of System	Transaction	Facility	Storage
Communications	Message	Switch	Trunk
Transportation	Car	Toll booth	Road
Data processing	Record	Key punch	Computer memory

A trunk means a cable consisting of many wires, which can carry several messages simultaneously and is therefore represented by a storage. A switch in this case is assumed to pass only one message at a time and is represented by a facility.

Figure 11-1 shows four block types, SEIZE, RELEASE, ENTER, and LEAVE, concerned with using facilities and storages. Field A in each case indicates which facility or storage is intended, and the choice is usually marked in the flag attached to the symbols of the blocks. The SEIZE block allows a transaction to engage a facility if it is available. The RELEASE block allows the transaction to disengage the facility. In an analogous manner, an ENTER block allows a transaction to occupy space in a storage, if it is available, and the LEAVE block allows it to give up the space. If the fields B of the ENTER and LEAVE blocks are blank, the storage contents are changed by 1. If there is a number (≥ 1), then the contents change by that value. Any number of blocks may be placed between the points at which a facility is seized and released to simulate the actions that would be taken while a transaction has control of a facility. Similar arrangements apply for making use of storages.

To illustrate the use of these block types, consider again the manufacturing shop discussed before. Since the average inspection time is 4 minutes and the average generation rate of parts is one every 5 minutes, there will normally be only one part inspected at a time. Occasionally, however, a new part can arrive before the previous part has completed its inspection. With an ADVANCE block representing inspection, this situation will result in more than one transaction being at the ADVANCE block at one time, and the simulation must be interpreted as meaning that more than one inspector is available.

Assuming that there is only one inspector, it is necessary to represent the inspector by a facility, to simulate the fact that only one part at a time can be inspected. The block diagram will then appear as shown in Fig. 11-5. A SEIZE block and a RELEASE block have been added to simulate the engaging and disengaging of the inspector. Coding and results for this model are given in Fig. 11-6. A line of output is given for each facility, showing how many times

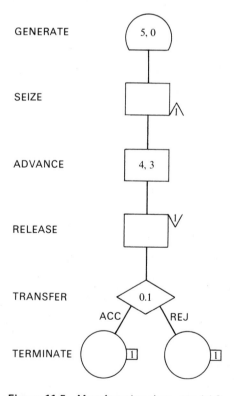

Figure 11-5. Manufacturing shop—model 2.

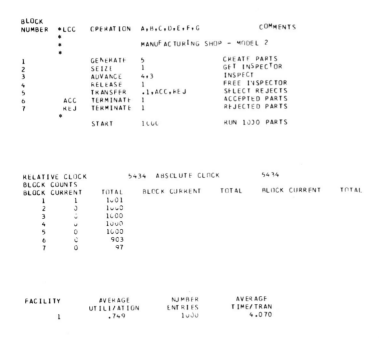

```
BLOCK
NUMBER  *LOC    OPERATION  A,B,C,D,E,F,G              COMMENTS
         *
         *
         *                 MANUFACTURING SHOP - MODEL 2

  1              GENERATE   5                CREATE PARTS
  2              SEIZE      1                GET INSPECTOR
  3              ADVANCE    4,3              INSPECT
  4              RELEASE    1                FREE INSPECTOR
  5              TRANSFER   .1,ACC,REJ       SELECT REJECTS
  6      ACC     TERMINATE  1                ACCEPTED PARTS
  7      REJ     TERMINATE  1                REJECTED PARTS
         *
                 START      1000             RUN 1000 PARTS



RELATIVE CLOCK          5434   ABSOLUTE CLOCK        5434
BLOCK COUNTS
BLOCK CURRENT   TOTAL    BLOCK CURRENT   TOTAL   BLOCK CURRENT   TOTAL
  1       1     1001
  2       0     1000
  3       0     1000
  4       0     1000
  5       0     1000
  6       0      903
  7       0       97



FACILITY        AVERAGE         NUMBER      AVERAGE
                UTILIZATION     ENTRIES     TIME/TRAN
       1          .749           1000        4.070
```

Figure 11-6. Coding and results for model 2.

it was seized, the average utilization made of the facility, and the average time transactions held the facility. In this case, the results show that the inspector was busy for 74.9% of his time.

If more than one inspector is available, they can be represented as a group by a storage with a capacity equal to the number of inspectors. The SEIZE and RELEASE blocks of Fig. 11-5 would then be replaced by an ENTER and a LEAVE block, respectively. Suppose, for example, the inspection time were three times as long as before; three inspectors would be justified and the model would be coded as listed in Fig. 11-7. Note that a control card called STORAGE has been added. In the location field, it identifies the storage, and the A field carries the capacity of the storage ($\leqslant 2,147,483,647$). Results for this model are also shown in Fig. 11-7. A line of output appears for each storage, giving the information indicated in the headings. It shows, for example, that, on the average, 2.275 of the inspectors were busy at any time.

The case of a single inspector, shown in Fig. 11-5, could have been modeled by using a storage with capacity 1 instead of a facility. An important logical difference between the two possible representations is that, for a facility, only the transaction that seized the facility can release it, whereas, for a storage, entering and leaving can be separate actions carried out independently by different transactions. No use was made of this property in these models but the difference is of importance in many models.

```
BLOCK
NUMBER    *LOC    OPERATION   A,B,C,D,E,F,G              COMMENTS
          *
          *                   MANUFACTURING SHOP - MODEL 3
          *
1                 GENERATE    5                 CREATE PARTS
2                 ENTER       1                 GET AN INSPECTOR
3                 ADVANCE     12,9              INSPECT
4                 LEAVE       1                 FREE INSPECTOR
5                 TRANSFER    .1,ACC,REJ        SELECT REJECTS
6         ACC     TERMINATE   1                 ACCEPTED PARTS
7         REJ     TERMINATE   1                 REJECTED PARTS
          *
          1       STORAGE     3                 NUMBER OF INSPECTORS
          *
                  START       1000              RUN 1000 PARTS
```

```
RELATIVE CLOCK          5346   ABSOLUTE CLOCK          5346
BLOCK COUNTS
BLOCK CURRENT   TOTAL     BLOCK CURRENT   TOTAL   BLOCK CURRENT   TOTAL
  1       1     1003
  2       0     1002
  3       2     1002
  4       0     1000
  5       0     1000
  6       0      896
  7       0      104
```

STORAGE	CAPACITY	AVERAGE CONTENTS	AVERAGE UTILIZATION	ENTRIES	AVERAGE TIME/TRAN	CURRENT CONTENTS	MAXIMUM CONTENTS
1	3	2.275	.758	1002	12.138	2	3

Figure 11-7. Coding and results for model 3.

11-7

Gathering Statistics

Certain block types in GPSS are constructed for the purpose of gathering statistics about the system performance, rather than to represent system actions. The QUEUE, DEPART, MARK, and TABULATE blocks shown in Fig. 11-1 serve this purpose. They introduce two other entities of the GPSS program, *queues* and *tables*. As with facilities and storages, there can be many such entities up to a prescribed limit (usually, 300 for queues and 100 for tables) and they are individually identified by a number or symbolic name.

When the conditions for advancing a transaction are not satisfied, several transactions may be kept waiting at a block. They are kept in order by the program and, when the conditions are favorable, are allowed to move on according to priority and usually by a first-in, first-out rule. No information about the queue of transactions is gathered, however, unless they have been entered into a queue entity. The QUEUE block increases and the DEPART block decreases queue number A. If field B is blank, the change is a unit change; otherwise the value of field B (≥ 1) is used. The program measures the average and maximum queue lengths and, if required, the distribution of time spent on the queue.

It is also desirable to measure the length of time taken by transactions to move through the system or parts of the system, and this can be done with the MARK and TABULATE blocks. Each of these block types notes the time a transaction arrives at the block. The MARK block simply notes the time of arrival on the transaction. (If field A is blank, a special word is used. With n in field A, the nth parameter is used.) The TABULATE block subtracts the time noted by a MARK block from the time of arrival at the TABULATE block. The time, referred to as *transit time*, is entered in a table whose number or name is indicated in field A of the TABULATE block. If the transaction entering a TABULATE block has not passed through a MARK block, the transit time is derived by using as a base the time at which the transaction was created. In effect, the transit time of a transaction can be regarded as the time the transaction has been in the system, and the action of the MARK block is to reset the transit time to zero.

To illustrate the use of these blocks, we return to the model of a manufacturing shop with three inspectors. The properties of the GENERATE block are such that, if a transaction is unable to leave the block at the time it is created, no further creations are made until the block is cleared. The situation simulated in Fig. 11-5 therefore assumes that if one part is finished while the previous part is being inspected, the machining of further parts stops until the machine is cleared.

A more realistic situation is that parts will accumulate on the inspector's work bench if inspection does not finish quickly enough. In that case, it would be of interest to measure the size of the queue of work that occurs. This can be done as shown in Fig. 11-8, which expands upon model 3. A QUEUE block using queue number 1 is placed immediately before the ENTER block, and a DEPART block is placed immediately after the ENTER block to remove the part from the queue when inspection begins. Any transaction that does not have to wait for an inspector to become available will move through the queue without delay. Those that must wait will do so in the QUEUE block, and the program will automatically measure the length of stay in the queue. Figure 11-8 also shows a MARK and a TABULATE block placed so that they will measure how long the parts take to be inspected, excluding their waiting time in the queue. If the MARK block is omitted, the tabulated time is the time since the transactions first entered the system.

Coding for this model, together with results, is shown in Fig. 11-9. Note that another control card called TABLE has been introduced. In the location field, it identifies a table; and, in the A field, it indicates the quantity to be tabulated. In the B, C, and D fields of the TABLE card are the lower limit of the table, the tabulation interval size, and the number of intervals, respectively. The definitions of these terms are the same as those introduced in Sec. 10-5 and illustrated in Fig. 10-3. The symbol M1 in field A of the TABLE card indicates transit time tabulation.

It will be seen in Fig. 11-9 that a line of output occurs for each queue, and a table is printed in accordance with the specifications of the TABLE card. The

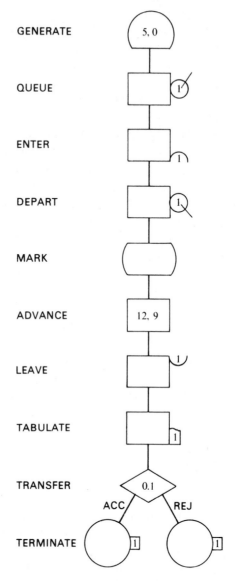

GENERATE 5, 0

QUEUE 1

ENTER 1

DEPART 1

MARK

ADVANCE 12, 9

LEAVE 1

TABULATE 1

TRANSFER 0.1

ACC   REJ

TERMINATE 1   1

Figure 11-8. Manufacturing shop-model 4.

TABULATE blocks can be used to gather a variety of other statistics, as will be explained in Sec. 12-2.

11-8

Conditional Transfers

As a last example in this chapter, we illustrate the use of both the conditional and unconditional transfer modes of the TRANSFER block. Consider again the case of three inspectors but suppose that the manufactured parts are put

```
BLOCK
NUMBER  *LOC    OPERATION  A,B,C,D,E,F,G           COMMENTS
         *
         *                 MANUFACTURING SHOP - MODEL 4
         *
  1              GENERATE   5                       CREATE PARTS
  2              QUEUE      1                       QUEUE FOR INSPECTOR
  3              ENTER      1                       GET AN INSPECTOR
  4              DEPART     1                       LEAVE QUEUE
  5              MARK
  6              ADVANCE    12,9                     INSPECT
  7              LEAVE      1                       FREE INSPECTOR
  8              TABULATE   1                       MEASURE TRANSIT TIME
  9              TRANSFER   .1,ACC,REJ               SELECT REJECTS
 10      ACC     TERMINATE  1                       ACCEPTED PARTS
 11      REJ     TERMINATE  1                       REJECTED PARTS
         *
         1       STORAGE    3                       NUMBER OF INSPECTORS
         1       TABLE      M1,5,5,10               TABULATION INTERVALS
         *
                 START      1000                    RUN 1000 PARTS
```

```
RELATIVE CLOCK        5016   ABSOLUTE CLOCK        5016
BLOCK COUNTS
BLOCK CURRENT   TOTAL    BLOCK CURRENT   TOTAL    BLOCK CURRENT   TOTAL
  1       0     1003       11      0      100
  2       1     1003
  3       0     1002
  4       0     1002
  5       0     1002
  6       2     1002
  7       0     1000
  8       0     1000
  9       0     1000
 10       0      900
```

STORAGE	CAPACITY	AVERAGE CONTENTS	AVERAGE UTILIZATION	ENTRIES	AVERAGE TIME/TRAN	CURRENT CONTENTS	MAXIMUM CONTENTS
1	3	2.430	.810	1002	12.165	2	3

QUEUE	MAXIMUM CONTENTS	AVERAGE CONTENTS	TOTAL ENTRIES	ZERO ENTRIES	PERCENT ZEROS	AVERAGE TIME/TRANS	$AVERAGE TIME/TRANS	TABLE NUMBER	CURRENT CONTENTS
1	2	.149	1003	742	73.9	.746	2.869		1

$AVERAGE TIME/TRANS = AVERAGE TIME/TRANS EXCLUDING ZERO ENTRIES

```
TABLE   1
ENTRIES IN TABLE         MEAN ARGUMENT      STANDARD DEVIATION      SUM OF ARGUMENTS
       1000                 12.167                5.371                12168.000         NON-WEIGHTED
```

UPPER LIMIT	OBSERVED FREQUENCY	PER CENT OF TOTAL	CUMULATIVE PERCENTAGE	CUMULATIVE REMAINDER	MULTIPLE OF MEAN	DEVIATION FROM MEAN
5	145	14.49	14.4	85.5	.410	-1.334
10	244	24.39	38.8	61.1	.821	-.403
15	300	29.99	68.8	31.1	1.232	.527
20	258	25.79	94.6	5.3	1.643	1.458
25	53	5.29	100.0	.0	2.054	2.389

REMAINING FREQUENCIES ARE ALL ZERO

Figure 11-9. Coding and results for model 4.

on a conveyor, which carries the parts past inspectors placed at intervals along the conveyor. It takes 2 minutes for a part to reach the first inspector; if he is free at the time the part arrives, he takes it for inspection. If he is busy at that time, the part takes a further 2 minutes to reach the second inspector, who will take the part if he is not busy. Parts that pass the second inspector may get

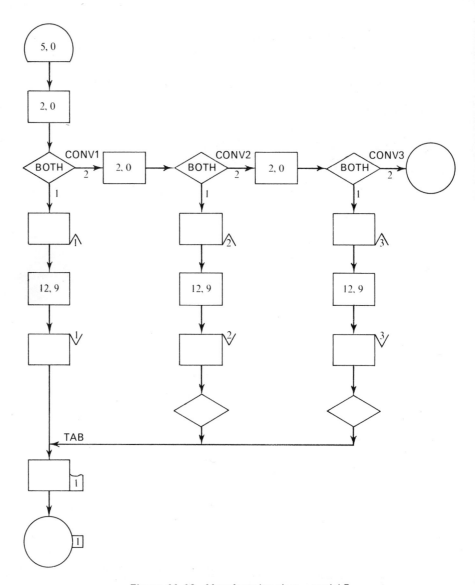

Figure 11-10. Manufacturing shop—model 5.

picked up by the third inspector, who is a further 2 minutes along the conveyor belt; otherwise, they are lost. To keep the model small, only the transit time of the parts will be recorded and the possibility of the inspectors rejecting parts will be ignored.

A block diagram for the system is shown in Fig. 11-10. The movement of parts along the conveyor is represented by **ADVANCE** blocks, each with an action time of 2 minutes. As a transaction leaves an **ADVANCE** block, it tests whether an inspector is available by entering a TRANSFER block with a selection factor set to BOTH. Exit 1 of each of these TRANSFER blocks leads to a SEIZE block representing one of the inspectors. If the inspector is free at the time the transaction enters, the transaction will leave by way of exit 1 to take over the facility. If the inspector is busy, the transaction will pass to the next stage of processing by way of exit 2.

When parts finish inspection, the transactions go to a single TABULATE block where the transit time is recorded. Because of the rule by which transactions normally pass from one block to the next higher numbered block, only one of the three RELEASE blocks that complete the inspection phase is able to pass the transactions directly to the TABULATE block. The others pass the transactions to TRANSFER blocks that send the transactions unconditionally to the TABULATE block. It would be possible to have three separate TABULATE blocks at the exit of each RELEASE block, making the unconditional TRANSFER blocks unnecessary. Each TABULATE block can refer to the same table number, so that the two alternative ways of drawing the block diagram will lead to exactly the same statistics. This principle—allowing the same GPSS entity to be referred to in more than one place—applies to all entities.

Coding and results for the problem are shown in Fig. 11-11. An explanation of the extra START card shown in the coding will be given in the next section. Note how the rule for normally moving transactions to successively numbered blocks causes the coding to be broken into several parts. Breaks are caused by both the unconditional TRANSFER and the conditional TRANSFER blocks. A way of simplifying block diagrams, such as this one, that have repetitive segments will be discussed in Sec. 12-8.

11-9

Program Control Cards

The first card of the GPSS input deck is a control card with the word SIMULATE in the operation field. Without this card, the problem will be assembled but not executed.

It is desirable to be able to stop and restart a simulation run and also repeat a simulation run with changes to some of the values in the model. When a GPSS simulation run is finished, therefore, the program does not immediately destroy the used model. Instead, it looks for more input cards following the START card, keeping the model exactly as it was at the completion of the run. Cards following the START card can change the model. For example, a storage capacity could be redefined by inserting a STORAGE card that refers to the number of a previously defined storage and gives the new value. The model could also be modified by changing existing blocks or adding new blocks. When all such modifications have been read and the model is ready for rerunning another START card will instruct the program to begin simulation again.

```
BLOCK
NUMBER  *LOC   OPERATION   A,B,C,D,E,F,G              COMMENTS
        *
        *                  MANUFACTURING SHOP - MODEL 5.
        *
  1            GENERATE    5                   CREATE PARTS
  2            ADVANCE     2                   MOVE TO 1ST INSPTOR
  3            TRANSFER    BOTH,,CONV1         CHECK IF HE IS BUSY
  4            SEIZE       1                   GET 1ST INSPECTOR
  5            ADVANCE     12,9                INSPECT
  6            RELEASE     1                   FREE INSPECTOR
  7     TAB    TABULATE    1                   MEASURE TRANSIT TIME
  8            TERMINATE   1
        *
  9     CONV1  ADVANCE     2                   MOVE TO 2ND INSPECTOR
 10            TRANSFER    BOTH,,CONV2         CHECK IF HE IS BUSY
 11            SEIZE       2                   GET 2ND INSPECTOR
 12            ADVANCE     12,9                INSPECT
 13            RELEASE     2                   FREE INSPECTOR
 14            TRANSFER    ,TAB                GO TO TABULATE BLOCK
        *
 15     CONV2  ADVANCE     2                   MOVE TO 3RD INSPECTOR
 16            TRANSFER    BOTH,,CONV3         CHECK IF HE IS BUSY
 17            SEIZE       3                   GET 3RD INSPECTOR
 18            ADVANCE     12,9                INSPECT
 19            RELEASE     3                   FREE INSPECTOR
 20            TRANSFER    ,TAB                GO TO TABULATE BLOCK
        *
 21     CONV3  TERMINATE
        *
        1      TABLE       M1,5,5,10           TABULATION INTERVALS
        *
               START       10,NP               INITIALIZE WITH 10 PARTS
               RESET                           WIPE OUT STATISTICS
               START       1000                RUN 1000 PARTS
```

```
RELATIVE CLOCK      5670  ABSOLUTE CLOCK      5750
BLOCK COUNTS
BLOCK CURRENT    TOTAL    BLOCK CURRENT   TOTAL    BLOCK CURRENT   TOTAL
   1      0      1134      11      0      347       21      0      135
   2      0      1134      12      1      347
   3      0      1134      13      0      347
   4      0       399      14      0      347
   5      0       399      15      1      388
   6      0       399      16      0      388
   7      0      1000      17      0      253
   8      0      1000      18      0      253
   9      0       735      19      0      254
  10      0       735      20      0      254
```

FACILITY	AVERAGE UTILIZATION	NUMBER ENTRIES	AVERAGE TIME/TRAN	SEIZING TRANS. NO.	PREEMPTING TRANS. NO.
1	.863	399	12.273		
2	.756	348	12.333	5	
3	.565	254	12.629		

```
TABLE    1
ENTRIES IN TABLE          MEAN ARGUMENT          STANDARD DEVIATION       SUM OF ARGUMENTS
     1000                    16.115                    5.648                 16116.000         NON-WEIGHTED
```

UPPER LIMIT	OBSERVED FREQUENCY	PER CENT OF TOTAL	CUMULATIVE PERCENTAGE	CUMULATIVE REMAINDER	MULTIPLE OF MEAN	DEVIATION FROM MEAN
5	30	2.99	2.9	97.0	.310	-1.967
10	169	16.89	19.8	80.1	.620	-1.082
15	259	25.89	45.7	54.2	.930	-.197
20	261	26.09	71.8	28.1	1.241	.687
25	258	25.79	97.6	2.3	1.551	1.572
30	23	2.29	100.0	.0	1.861	2.458

REMAINING FREQUENCIES ARE ALL ZERO

Figure 11-11. Coding and results for model 5.

Certain control cards can be included between START cards to set the conditions under which the rerun is made. A card with the single word RESET in the operation field will wipe out all the statistics gathered so far but will leave the system loaded with transactions. The purpose is to gather statistics in the second run with the initial build-up period excluded. The output of the first run is not then of interest and is usually suppressed by putting the letters NP in the B field of the START card. This procedure has been followed in the simulation shown in Fig. 11-11. One START card has run the model for a total of 10 completed transactions, and the output of this run has been suppressed. A RESET card has wiped out the statistics gathered in that run; the second START card has restarted the simulation from the point at which the tenth transaction finished and has continued for a further 1,000 transactions. The results shown are the statistics gathered during the second run.

The RESET card also sets to zero a relative clock. When the run is completed, the output marked ABSOLUTE CLOCK gives the time since the run began; the output marked RELATIVE CLOCK gives the time since the last RESET card. If no RESET card is used, the two times are the same.

Another control card, CLEAR, would not only wipe out the statistics of the preceding run, but would also wipe out the transactions in the system; so that, the rerun does in fact start the simulation from the beginning. This procedure would be followed when the objective is to repeat the problem with a change in some value. Although the CLEAR card returns the model to its initial state, it does not reset the random number generator seed. The following sequence of cards

START

CLEAR

START

would run the same problem twice but the second run would use a different set of random numbers. When there are stochastic processes in the model, the second run will be a second sample of the possible outputs that could arise. (It is possible to limit the range of the RESET and CLEAR actions, see (5).)

As well as controlling different runs of the same model, GPSS allows any number of different problems to be run successively with a single loading of the program. A control card with the word JOB in the operation field instructs the program to wipe out the entire model preceding the card and proceed with the following problem. This can be repeated as many times as desired. All five models described in this chapter were run at one time by placing a JOB card between the runs. A JOB card is not needed in front of the first problem. Unlike the CLEAR card, the JOB card does reset the random number seed so that each problem finds the program in exactly the same form as does the first problem.

Since the program always anticipates further input when it completes a run, a control card with the word END in the operation field must be placed at the end of all problems, even if there is only one, to terminate all simulation. The SIMULATE control card appears only once, at the beginning of the input deck, even if there are multiple jobs.

Exercises

Give GPSS block diagrams and write programs for the following problems:

11-1   In the manufacturing shop model of Fig. 11-2, suppose that parts rejected by the inspector are sent back for further work. Reworking takes 15 ± 3 minutes and does not involve the machine that originally made the parts. After correction, the parts are re-submitted for inspection. Simulate for 1,000 parts to be completed.

11-2   Parts are being made at the rate of one every 6 minutes. They are of two types, A and B, and are mixed randomly with about 10% being type B. A separate inspector is assigned to examine each type of part. The inspection of A parts takes 4 ± 2 minutes and B parts take 20 ± 10 minutes. Both inspectors reject about 10% of the parts they inspect. Simulate for a total of 1,000 type A parts accepted.

11-3   Workers come to a supply store at the rate of one every 5 ± 2 minutes. Their requisitions are processed by one of two clerks who take 8 ± 4 minutes for each requisition. The requisitions are then passed to a single storekeeper who fills them one at a time, taking 4 ± 3 minutes for each request. Simulate the queue of workers and measure the distribution of time taken for 1,000 requisitions to be filled.

11-4   People arrive at an exhibition at the rate of one every 3 ± 2 minutes. There are four galleries. All visitors go to gallery A. Eighty percent then go to gallery B, the remainder go to gallery C. All visitors to gallery C move on to gallery D and then leave. None of the visitors to gallery B goes to gallery C; however, about 10% of them do go to gallery D before leaving. Tabulate the distribution of the time it takes 1,000 visitors to pass through when the average times spent in the galleries are as follows:

$$A, 15 \pm 5; \quad B, 30 \pm 10; \quad C, 20 \pm 10; \quad D, 15 \pm 5$$

11-5 People arrive at the rate of one every 10 ± 5 minutes to use a single telephone. If the telephone is busy, 50% of the people come back 5 minutes later to try again. The rest give up. Assuming a call takes 6 ± 3 minutes, count how many people will have given up by the time 1,000 calls have been completed.

11-6 People arrive at a cafeteria with an inter-arrival time of 10 ± 5 seconds. There are two serving areas, one for hot food and the other for sandwiches. The hot food area is selected by 80% of the customers and it has 6 servers. The sandwich area has only one server. Hot food takes 1 minute to serve and sandwiches take $\frac{1}{2}$ minute. When they have been served, the customers move into the cafeteria which has seating capacity for 200 people. The average time to eat a hot meal is 30 ± 10 minutes and for a sandwich is 15 ± 5 minutes. Measure the queues for service and the distribution of time to finish eating from the time of arrival. Simulate for 1,000 people to finish their meal.

11-7 Ships arrive at a harbor at the rate of one every $1 \pm \frac{1}{2}$ hours. There are six berths to accommodate them. They also need the services of a crane for unloading and there are five cranes. After unloading, 10% of the ships stay to refuel before leaving; the others leave immediately. Ships do not need the cranes for refueling. Simulate the queues for berths and cranes assuming it takes $7\frac{1}{2} \pm 3$ hours to unload and $1 \pm \frac{1}{2}$ hours to refuel. Simulate for 100 ships to clear the harbor.

Bibliography

1 Gordon, Geoffrey, "A General Purpose Systems Simulation Program," *Proc. of EJCC, Washington, D.C.*, New York: The Macmillan Company, 1961, 87–104.

2 Efron, R., and G. Gordon, "A General Purpose Digital Simulator and Examples of its Application: Part 1—Description of the Simulator," *IBM Systems Journal*, III, No. 1 (1964), 21–34.

3 Herscovitch, H., and T. Schneider, "GPSS III—An Expanded General Purpose Simulator," *IBM Systems Journal*, IV, No. 3 (1965), 174–183.

4 GPSS/360 Introductory User's Manual, IBM Corp., Form No. H 20-0304.

5 GPSS/360 User's Manual, IBM Corp., Form No. H 20-0326.

12

GPSS EXAMPLES

The previous chapter introduced the concepts of the GPSS program, described some of the block types, and illustrated the manner in which the program is used. In this chapter, a fuller account of the program will be given together with descriptions of more block types. First, a more complete description of transactions will be given. (See Fig. 11-1 for the block symbols, Table 11-1 for coding information for the blocks, and Table 11-2 for control cards.)

As described so far, transactions have no particular identity; each is treated by a block in the same manner as any other transaction. In fact, transactions have two types of attribute which influence the way they are processed. Each transaction has one of 128 levels of *priority*, indicated by the numbers 0 to 127, with 0 being the lowest priority. At any point in the block diagram, the priority can be set up or down to any of the levels by the PRIORITY block. The block is coded by putting the priority in field A of the block. It is also possible to designate the priority at the time a transaction is created by putting the priority in the E field of the GENERATE block creating the transaction. If the field is left blank, the priority is set to 0.

When there is competition between transactions to occupy a block, the service rule is to advance transactions in order of priority and first-in, first-out within priority class. The program has the capability of implementing more complex queuing disciplines through the chain feature described in Sec. 12-10.

A transaction can also have *parameters* which are signed integral numbers. The number of parameters attached to a transaction can be from 0 to 100, and is specified in the F field of the GENERATE block creating the transaction. If the field is left blank, 12 parameters are automatically allocated.

All parameter values are zero at the time a transaction is created. A value is given to a parameter when a transaction enters the ASSIGN block. The ASSIGN block can use as a source of the value to be assigned any of the standard numerical attributes (SNA's) described in the next section. It is coded by indicating the number of the parameter in field A and the SNA to be used in field B. An ASSIGN block can either add to, subtract from, or replace the value of a parameter. A + or − sign immediately following the parameter number in field A indicates that the assigned value is to be added or subtracted. For replacement, the parameter number only is given.

Usually, parameters record characteristics of the entity represented by the transaction. For example, a transaction representing a message in a communication system might use parameters to represent the message length, the message type, or the destination. A transaction representing a ship might use parameters for carrying the cruising speed, number of passengers, or type of cargo. Another use of parameters is for logical control of the model. A transaction might, for example, be controlling the number of times an operation is to be performed with parameters carrying the number of times the operation has been performed so far and the total number required.

12-2

Standard Numerical Attributes

Parameters are items of data that represent attributes of the transaction. In addition, there are attributes of the other entities of the system, such as the number of transactions in a storage or the length of a queue, which are made available to the program user. Collectively, they are called *standard numerical attributes (SNA's)*. Each type of SNA is identified by a one or two letter code and a number. For example, the contents of storage number 5 is denoted by S5 and the length of queue number 15 is Q15. For completeness, parameters are included in the category of SNA's. They are identified by the symbol Pn, where n is the number of the parameter.

Computations can be carried out by defining *variable statements* in which simple SNA's are combined mathematically with the operators +, −, ★, /, and @ for addition, subtraction, multiplication, division, and modulo division, that is, deriving the remainder after division. Parentheses, up to five levels, may be included in the definition of a variable statement. A variable statement is numbered and defined by a card that has the number, or name, in the location field and the word VARIABLE in the operation field. The statement begins in column 19 and continues without blanks up to column 71 if necessary. For example, the following statement:

$$5 \text{ VARIABLE} \qquad S6 + 5 ★ (Q12 + Q17)$$

defines variable 5 as being the sum of the current contents of storage number 6 plus 5 times the sum of the lengths of queues 12 and 17. Variable statements are included in the category of SNA's and denoted by the symbol Vn, so that one variable statement can be a component of another.

Many uses can be made of SNA's. They provide the inputs to the functions described in the next section, thereby allowing a great variety of functional relationships to be introduced into the model. It will be recalled that the TABLE card used in conjunction with the TABULATE block to measure transit time carried M1 in field A. Transit time is one of the SNA's and M1 is its symbol; any other SNA can be used in a TABLE card, allowing the program to collect and tabulate a wide variety of statistics. The current value of any SNA may also be used as the value of almost any field of a block. Exceptions are fields that contain program key-words, such as the selection factor field of the TRANSFER block.

In general, the values of the SNA's change as the simulation proceeds. The program does not keep the current value at all times; it computes the value of SNA's at the time they are needed. It is convenient to be able to save values computed at one time for use at some later time, and this can be done by the use of a block type called SAVEVALUE. The block indicates in field A the number of one of many *savevalue locations* and, in field B, gives the SNA to be

Table 12-1 GPSS Standard Numerical Attributes

C1   The current value of clock time.

CHn   The number of transactions on chain n.

Fn   The current status of facility number n. This variable is 1 if the facility is busy and 0 if not.

FNn   The value of function n. (The function value may be computed to have a fractional part but the SNA gives only the integral part, unrounded.)

Kn   The integer n (the notation n may also be used).

M1   The transit time of a transaction.

Nn   The total number of transactions that have entered block n.

Pn   Parameter number n of a transaction.

Qn   The length of queue n.

Rn   The space remaining in storage n.

RNn   A computed random number having one of the values 1 through 999 with equal probability. (When the reference is made to provide the input for a function, the value is automatically scaled to the range 0 to 1.) Eight different generators can be referenced by n = 1, 2, ..., 8.

Sn   The current occupancy of storage n.

Vn   The value of variable statement number n.

Wn   The number of transactions currently at block n.

Xn   The value of savevalue location n.

saved. As with an ASSIGN block, a + or − sign immediately following the savevalue number will result in the SNA value being added to or subtracted from the savevalue contents; otherwise, the value is replaced. The current contents of a savevalue location n is available as an SNA indicated by Xn. A special control card, called INITIAL, can set the value of a savevalue at the beginning of a simulation so that savevalues can introduce initial conditions. The notation Xn is put in the A field and the initial value is put in field B. The values of the non-zero savevalues are printed out at the end of the simulation run so that savevalues also provide a way of extracting results from the simulation other than through the standard report.

A partial list of the program SNA's is given in Table 12-1. The letter n in Table 12-1 is to be replaced by the number of some entity. The program allows any entity (but not a parameter) to be represented by a symbolic name. If a symbolic name is used, a reference to the entity in an SNA must use the name preceded by a $ sign. For example, a storage called BSKT would be referred to as S$BSKT.

12-3

Functions

To introduce functional relationships in a model, a GPSS model can include a number of functions. Each function is defined by giving two or more pairs of numbers that relate an input x to an output y. The function can be in a continuous or discrete mode. In a continuous mode, the program will interpolate linearly between the defined points, allowing the user to approximate a continuous function by a series of straight line segments. In a discontinuous mode, the function is treated as a "staircase function." If x_i and x_{i+1} are two successive points at which the function has been defined an input value in the range $x_i < x \leqslant x_{i+1}$ will result in the value y_i being produced. Figure 12-1 illustrates these two modes of using functions.

Any number of points (>1) can be used to define a function; and the size of the intervals between successive points can vary freely, except that a function

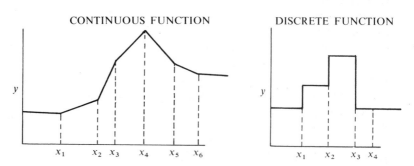

Figure 12-1. Continuous and discrete functions.

defined to have a random number as input must have values of x in the range 0 to 1. The values of x and y can be fractional and negative. A reference to a value of x below the lowest defined value, x_1, results in the value y_1. Similarly, reference beyond the highest defined value of x produces the value of y at that highest value.

In using a function, any of the SNA's can be defined as being the input. Examples of some of the more commonly used choices of input and their uses are:

(a) By using a uniformly distributed random number RNn, as input, any other distribution can be obtained with the technique of Sec. 6-9.
(b) By using clock time C1, action times can be made to depend upon clock time, thereby simulating the effects of peak loads.
(c) By using the current contents of a storage, an action time can depend upon the current load on some part of the system.
(d) By using a parameter Pn, an action time can depend on each particular transaction.

The choice of input is made at the time of defining the function. It is *not* selected when the function is referenced. At least two cards are needed to define a function. The first is called a FUNCTION card and it is immediately followed by one or more *function data* cards. The FUNCTION card is punched to the following format:

Field	Contents
Location	Function number
Operation	FUNCTION
A	SNA to be used as input
B	Cn for continuous mode
	Dn for discontinuous mode,
	where n is the number of points
	to be defined.

The function data cards carry the values of x and y that define the function. Values are punched beginning in column 1 with commas between the x and y values and slashes between successive pairs thus: $x_1,y_1/x_2,y_2/,\ldots$. Any number of points may be on one card as long as punching does not go beyond column 71. Any number of additional cards may be used but no one pair of x, y values may be split between two cards. The listing in Fig. 12-4 shows, as function number 1, the coding for the approximation to the continuous function

$$y = \text{Log}_e (1 - x)$$

that was given previously in Table 7-1. It also shows, as function number 2, the coding for a discontinuous function to the following specification:

x	y
$0 < x \leqslant 0.2$	5
$0.2 < x \leqslant 0.5$	10
$0.5 < x \leqslant 0.9$	15
$0.9 < x \leqslant 1.0$	20

12-4

Transfer Modes   A further important use of both parameters and functions is in conjunction with the TRANSFER block. Rather than just making a random choice or a conditional selection between two blocks, a TRANSFER block can use the value of a parameter or a function as the location to which it sends a transaction. To use the parameter mode, a P is put in the A field of the TRANSFER block and the parameter number is put in field B. For the function mode, FN is put in field A and the function number is put in field B. If a parameter number is put in field C, the parameter value is added to the function value. Both modes of operation are particularly useful when different categories of transactions must be handled in the same way in one part of the block diagram but, eventually, must be separated.

Continuous or discrete functions using any SNA as input can be employed, but a particular mode of function has been defined to simplify the use of functions for transfers. A *list-mode* function assumes that the input is an integer n, and it returns the nth listed value of the function. Since location numbers are not always known before assembly, it is permissible to use location names as function values; the assembly program will supply the correct numerical value. Suppose, for example, transactions are to be sent to one of four locations called LOCA, LOCB, LOCC, and LOCD, and suppose further that one of the numbers 1, 2, 3, or 4 has been placed in a parameter, say number 3. The following coding will effect the transfer:

```
                    TRANSFER     FN,1

                      . . . . . .

       1            FUNCTION     P3,L4

                1,LOCA/2,LOCB/3,LOCC/4,LOCD
```

The characters L4 signify that the function is in the list mode and has four listed values.

Alternatively, an ASSIGN block could have used this function to set a parameter, say number 1, to be the location. Later, the transfer could be effected with the block

TRANSFER   P,1

Table 12-2 summarizes all the modes of the TRANSFER block that have been described.

Table 12-2   GPSS Transfer Modes

Mode	Field A	Field B	Field C
Unconditional	*0* (blank)	Next Block *A*	
Random	*.xxx*	Next Block *A*	Next Block *B*
Conditional	BOTH	Next Block *A*	Next Block *B*
Parameter	P	Parameter No.	—
Function	FN	Function No.	Parameter No.

12-5
Simulation of a Supermarket

To illustrate the use of functions, parameters, and SNA's, a simulation model will be written for a supermarket that operates in the following manner. Customers of the supermarket are obliged to take a basket before they begin to shop. There is a limited number of baskets and, if no basket is available when they arrive, customers leave without shopping. If they get a basket, customers shop and then check-out at one of five check-out counters. After checking-out, they return the baskets and leave the supermarket. A block diagram of the model is shown in Fig. 12-2 and the GPSS coding is shown in Fig. 12-4. There are four sections concerned with

(a) Getting a basket
(b) Shopping
(c) Checking-out
(d) Leaving

Each shopper is represented by a transaction and the unit of time is 1 second.

A GENERATE block creates the transactions that represent the customers. It is assumed that the arrival pattern can be represented by a Poisson distribution. The process for generating such numbers requires the function $y = \text{Log}_e (1 - x)$ which is coded as function number 1. (See Sec. 7-4.) The rule for using functions at GENERATE blocks is that the mean multiplies the function. In this case, a mean inter-arrival time of 36 will be assumed.

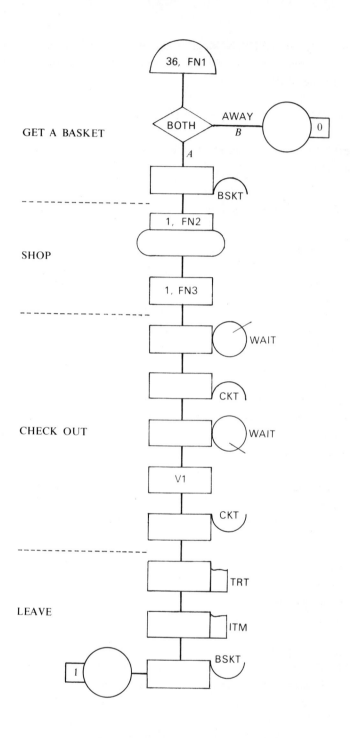

Figure 12-2. Simulation of a supermarket.

To represent the baskets, a storage denoted by BSKT is used with capacity equal to the number of baskets; in this case, there will be 50 baskets. The decision whether a customer is able to get a basket is made at a TRANSFER block immediately following the GENERATE block. The TRANSFER block has a BOTH selection factor and it attempts to send transactions to an ENTER block using the storage BSKT. If a basket is available, the ENTER block will accept the transaction and increase the count of baskets in use by 1. If the storage is full, however, no basket is available, and the TRANSFER block sends the transaction to a TERMINATE block called AWAY which counts the customers turned away for lack of baskets. Coding for this section is

```
        GENERATE    36,FN1
        TRANSFER    BOTH,,AWAY
        ENTER       BSKT

        . . . . . . . . . .
AWAY    TERMINATE
```

The simulation will be arranged to make shopping time depend upon the number of items purchased. A parameter, number 1, is assigned to represent the number of items to be purchased. The number of items is determined by an ASSIGN block using the discrete distribution of function number 2. Following the technique described in Sec. 6-8, a random number input to the function will result in the number of items being 5, 10, 15, or 20 with probabilities 0.2, 0.3, 0.4, and 0.1, respectively. Transactions then pass to an ADVANCE block to represent the shopping. The functional relationship shown in Fig. 12-3 and

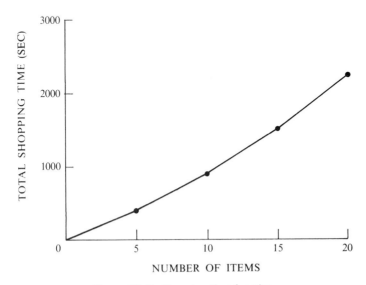

Figure 12-3. Shopping-time function.

coded as function number 3 is assumed between the number of items and shopping time. The function has parameter number 1 as input. The program evaluates the function with parameter 1 of each transaction that enters the ADVANCE block. The mean at the ADVANCE block is set to 1 so that the value of the function is applied directly as the action time. Coding for the shopping section is as follows:

```
ASSIGN    1,FN2
ADVANCE   1,FN3
```

When transactions emerge from the ADVANCE block, shopping is completed and they move to the section concerned with checking-out. There are five counters but it is not necessary in this study to distinguish the performance of the individual counters. The counters are therefore regarded as a service unit with five servers. They are represented by a storage, named CKT, that has a capacity of five.

There will be some congestion at the counters, and one of the objectives will be to measure the amount of congestion. The transactions therefore go to a QUEUE block, which enters them in a queue called WAIT. When a counter becomes available, a transaction leaves the QUEUE block for the ENTER block and then immediately goes to a DEPART block to be removed from the queue. Should there be no congestion, a transaction will move straight through the QUEUE, ENTER, and DEPART blocks.

Checking-out will be assumed to require 10 seconds to pay for each item plus 25 seconds for packing. Since parameter 1 is the number of items, the following variable statement will compute the checking-out time:

```
1         VARIABLE    P1★10+25
```

As mentioned before, an SNA can be placed in most fields of a block. In this case, field A of an ADVANCE block is set to be V1. As each transaction enters the block, the program computes the action time from variable statement number 1. When checking-out is completed, the transactions go to a LEAVE block to give up the counter space. The coding for the check-out section is

```
QUEUE       WAIT
ENTER       CKT
DEPART      WAIT
ADVANCE     V1
LEAVE       CKT
```

Upon completion of check-out, transactions pass to the section concerned with leaving the supermarket. They first go to a TABULATE block where the transit time is tabulated in a table called TRT. Suppose a record is also to be

```
BLOCK
NUMBER  *LOC   OPERATION  A,B,C,D,E,F,G                COMMENTS
        *      SIMULATICN OF A SUPERMARKET
        *
        1      FUNCTION   RN1,C24           FUNCTION FOR I/A INTERVAL
        0.0,0.0/0.1,0.104/C.2,0.222/0.3,0.355/0.4,0.509/0.5,0.69
        0.6,0.915/0.7,1.2/C.75,1.38/0.8,1.6/0.84,1.83/0.88,2.12
        0.9,2.3/0.92,2.52/C.94,2.81/0.95,2.99/C.96,3.2/.97,3.5
        0.98,3.9/C.99,4.6/C.995,5.3/C.998,6.2/0.999,7/0.9997,8
        *
        1      GENERATE   36,FN1            CREATE SHOPPERS
        2      TRANSFER   BOTH,,AWAY        CHECK FOR AVAILABLE BASKET
        3      ENTER      BSKT              GET A BASKET
        4      ASSIGN     1,FN2             DETERMINE NO. OF ITEMS
        5      ADVANCE    1,FN3             SHOP
        6      QUEUE      WAIT              WAIT FOR CCUNTER SPACE
        7      ENTER      CKT               GET COUNTER SPACE
        8      DEPART     WAIT              LEAVE QUEUE
        9      ADVANCE    V1                CHECK-OUT
        10     LEAVE      CKT               FREE COUNTER SPACE
        11     TABULATE   TRT               TABULATE TRANSIT TIME
        12     TABULATE   ITM               TABULATE NO. OF ITEMS
        13     LEAVE      BSKT              RETURN BASKET
        14     TERMINATE  1
               *
        15  AWAY TERMINATE                  LOST CUSTOMERS
               *
            TRT  TABLE     M1,500,500,10    TRANSIT TIME TABLE
            ITM  TABLE     P1,5,5,4         ITEM COUNT TABLE
               *
            CKT  STCRAGE   5                NUMBER OF COUNTERS
            BSKT STORAGE   50               NUMBER OF BASKETS
               *
        2      FUNCTION   RN1,D4            DISTR. OF NO. OF ITEMS
        .2,5/.5,10/.9,15/1.0,20
               *
        3      FUNCTICN   P1,C5             SHOPPING TIME DISTR.
        C,0/5,400/10,900/15,1500/20,2250
               *
        1      VARIABLE   P1*10€25          CHECK-OUT TIME
               *
               START      50,NP            INITIALIZE, SUPRESS PRINT
               RESET                       WIPE OUT STATISTICS
               START      1000             MAIN RUN
```

Figure 12-4. Coding of supermarket simulation.

made of the number of items bought by each customer. This is done by going to another TABULATE block which tabulates P1 in a table called ITM. The tabulation should, of course, reproduce the original distribution of function 2. This step has been inserted to illustrate the use of a TABULATE block for statistics other than transit time.

When tabulation is complete, transactions go to a LEAVE block, naming the storage BSKT, to represent the return of the basket. Finally they go to a TERMINATE block. Coding for this section is

```
        TABULATE     TRT
        TABULATE     ITM
        LEAVE        BSKT
        TERMINATE    1
TRT     TABLE        M1,500,500,10
ITM     TABLE        P1,5,5,4
```

Note that a 1 appears in the A field of this TERMINATE block while the A field of the other TERMINATE block is blank. The simulation run, therefore, will count only satisfied customers.

Coding for the complete model is shown in Fig. 12-4. Note a restraint on the order in which cards must be loaded in this example. Because a reference is made to a function by the GENERATE block, the function must be defined ahead of the GENERATE block.

12-6

Logic Switches

Two types of entities, facilities and storages, have been introduced to represent equipment. In addition, a third type of entity called *logic switches* is made available to represent two-state conditions in a system. Each switch is either on or off, and a block type called LOGIC is used to change the status of the switch. A transaction entering the block can either set the switch on, reset the switch off, or invert the switch from its current state. Should the switch already be in the desired state, no action is taken. In coding the LOGIC block, the letter S, R, or I indicating set, reset, or invert appears in *column 14* while the switch number, or name, appears in field A. The program keeps no statistics about logic switches. However, it prints out the numbers of the switches that are set at the time the run ends.

Examples of how logic switches are used are : a system representing a factory might use a logic switch to indicate whether a machine is in working order or not ; in a vehicular traffic system, a logic switch might represent a traffic light ; and, in a bank, a logic switch might represent whether a teller's position is open.

12-7

Testing Conditions

It is often desirable to control the flow of transactions according to prevailing conditions in the system. A block type called GATE can be used for this purpose. It can test the condition of any facility, storage, or logic switch in the block diagram. The conditions that can be tested and the symbols used to indicate the selected conditions are

LS	n	Logic switch n set
LR	n	Logic switch n reset
U	n	Facility n in use
NU	n	Facility n not in use
SF	n	Storage n full
SNF	n	Storage n not full
SE	n	Storage n empty
SNE	n	Storage n not empty

A transaction will enter the GATE block if the condition being tested is true. When the condition is not true, there is a choice of action. If an alternative block is specified, the transaction will be sent to the alternative block immediately. If no alternative is specified, the transaction will wait until the tested condition becomes true and then enter the GATE block. It is not necessary for the user to arrange for retesting; the program will automatically recognize when the condition changes and move the transaction at that time. In coding GATE blocks, the condition code begins from *column 13*, while the number or name of the entity to be tested is placed in field A. If an alternative block is specified, it is put in field B.

Another block type called TEST can test a variety of relationships between any two SNA's. Since the SNA's include variable statements, the TEST block is able to perform complex tests of conditions. The relationships that can be tested and the symbols used to represent them are shown below:

G	Greater than
GE	Greater than or equal
E	Equal
NE	Not equal
LE	Less than or equal
L	Less than

The relationship symbol begins in *column 13* and the two SNA's to be related are placed in fields A and B. Thus, the following block will test whether the content of storage 6 is less than the number recorded in savevalue location 12

```
TEST L      S6,X12
```

The action taken upon testing the condition is the same as for a GATE block. A transaction will enter the block if the condition is true and will either go to an alternative block if one is specified (in field C) or wait to enter when the condition changes.

12-8
Indirect Addressing

When reference is made to a field or SNA, the program must be supplied a specific number. Usually, the number is provided when the blocks are coded. However, the value can be left unspecified at that time, and the program can be arranged to take the value of a parameter of the transaction that enters a block or invokes the use of the SNA. This feature is referred to as *indirect addressing*. The notation ★n in a block field indicates that the value of parameter number *n* is to be used in that field. Since an SNA can be used in a block field, also, the notation Pn would achieve exactly the same effect but the notation ★n will be executed more quickly. Thus, in the supermarket example, the variable statement controlling the check-out time was written:

$$1 \qquad \text{VARIABLE} \qquad \text{P1}★10+25$$

It could also have been written

$$1 \qquad \text{VARIABLE} \qquad ★1★10+25$$

In evaluating this last form of the variable statement, the program will replace ★1 by the value of parameter number 1.

Indirect addressing can considerably reduce the size of a block diagram, where there are repetitive sections that differ only in the number of the entities used in the sections. A common section can be used, with parameters of the transactions supplying particular values. As an example, suppose that in the

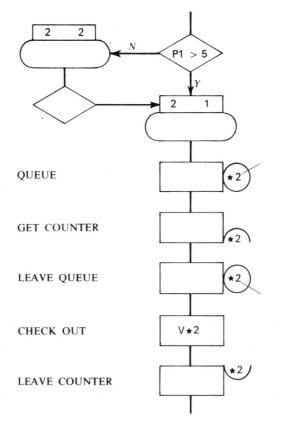

QUEUE

GET COUNTER

LEAVE QUEUE

CHECK OUT

LEAVE COUNTER

Figure 12-5. Indirect addressing version of check-out section.

supermarket problem, customers buying five or less items of shopping are allowed to use a special express check-out section of two counters. A new queue and storage are defined to represent the express area. The numbers of the queue and storage for the normal shoppers will both be 1 and the new queue and storage will both be number 2.

The section of the original solution concerned with the checking-out is now replaced with the block diagram shown in Fig. 12-5. Transactions go to a TEST block which tests whether parameter number 1, the number of shopping items, is greater than 5. If so, the transaction goes to an ASSIGN block which sets parameter number 2 to 1. Otherwise, the transaction goes to an ASSIGN block that sets parameter number 2 to 2. The other blocks of the section are the same as before, except that they have field A set to ★2. In each case, the entering transactions will cause the blocks to operate with the entity number indicated by the value of parameter number 2, so that the normal shoppers will join queue number 1 and use storage number 1, while the express shoppers will use number 2 in both cases.

On the assumption that different check-out times should be used for the two cases, a second variable statement, number 2, is defined for the express check-out time and the ADVANCE block representing the check-out time has the A field set to V★2. This makes the block select variable statement number 1 or 2 according to the value of parameter number 2.

12-9

GPSS Model of a Simple Telephone System

To illustrate the use of logic switches, a GPSS model of the telephone system discussed in Chaps. 8 and 9 will be derived. The system is one in which a series of calls come from a number of telephone lines and the system is to connect the calls by using one of a limited number of links. Only one call can be made to any one line at a time and it is assumed that calls are lost if the called party is busy or no link is available. Each line is represented by a logic switch whose number is the line number. The line is considered busy if the switch is set.

Each call is represented by one transaction; the unit of time chosen is 1 second. It will be assumed that the distribution of arrivals is Poisson with a mean inter-arrival time of 12 seconds. The length of the calls will also be assumed to have an exponential distribution. As in the previous examination of this system, it will be assumed that each new call can come from any of the non-busy lines with equal probability, and that its destination is equally likely to be any line other than itself.

A GPSS block diagram is shown in Fig. 12-6 and the coding in Fig. 12-7. A GENERATE block is used to create a series of transactions representing calls. The modifier at the block is the same function, number 1, used in the super-market example. The mean of the GENERATE block is set to 12. Parameters

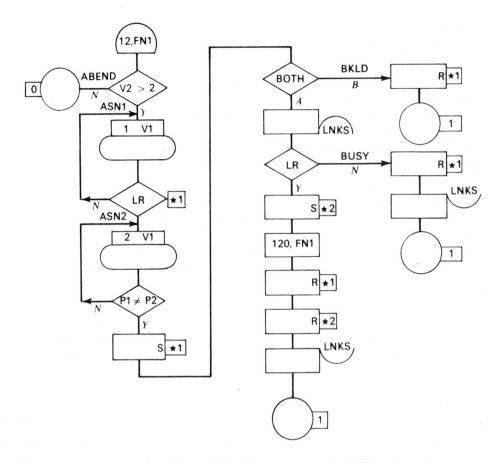

Figure 12-6. Telephone system in GPSS.

1 and 2 will be used to carry the origin and destination of the call. Each transaction is sent to two ASSIGN blocks to select and record the values. The source of the information is a VARIABLE statement, number 1, which will select a line at random by the method described in Sec. 6-8. The number of lines N is multiplied by a random number between 0 and 1, and the integral part is taken to represent a choice of 1 out of N. The value zero is not allowed for any GPSS entity, so 1 is added to the result to make the choice range from 1 to N. It is not necessary to take any action to extract the integral part because (with certain exceptions) GPSS works with integral numbers. Any evaluation of a VARIABLE statement or function that results in a fractional number is converted to an integer by dropping the fractional part. The number of lines will be varied on different runs so the desired number of lines is placed in savevalue number 1. An INITIAL card loads the desired value at the beginning of the program.

```
BLOCK
NUMBER   *LOC   OPERATION  A,B,C,D,E,F,G              COMMENTS
         *
         *     SIMULATION OF TELEPHONE SYSTEM - MODEL 1
         *
     1         FUNCTION   RN1,C24                FUNCTION FOR I/A INTERVAL
     0.0,0.0/0.1,0.104/0.2,C.222/C.3,0.355/0.4,0.5C9/0.5,0.69
     0.6,0.915/0.7,1.2/C.75,1.38/C.8,1.6/0.84,1.83/0.88,2.12
     0.9,2.3/0.92,2.52/0.94,2.81/C.95,2.99/0.96,3.2/.97,3.5
     0.98,3.9/0.99,4.6/0.995,5.3/C.998,6.2/0.999,7/0.9997,8
         *
     1         GENERATE   12,FN1                 CREATE CALLS
     2         TEST G     V2,2,ABND              TEST IF SYSTEM IS FULL
     3   ASN1  ASSIGN     1,V1                   PICK ORIGIN
     4         GATE LR    *1,ASN1                TEST FOR BUSY
     5   ASN2  ASSIGN     2,V1                   PICK DESTINATION
     6         TEST NE    P1,P2,ASN2             RETRY IF DEST = ORIGIN
     7         LOGIC S    *1                     MAKE ORIGIN BUSY
     8         TRANSFER   BOTH,,BLKD             TRY FOR LINK
     9   GETL  ENTER      LNKS                   GET LINK
    10         GATE LR    *2,BUSY                TEST FOR BUSY
    11         LOGIC S    *2                     MAKE DEST. BUSY
    12         ADVANCE    120,FN1                TALK
    13         LOGIC R    *1                     ORIGIN HANGS UP
    14         LOGIC R    *2                     DEST. HANGS UP
    15         LEAVE      LNKS                   FREE LINK
    16   TERM  TERMINATE  1
         *
    17   ABND  TERMINATE                         ABANDON CALL
         *
    18   BLKD  LOGIC R    *1                     ORIGIN HANGS UP
    19         TERMINATE  1                      BLOCKED CALLS
         *
    2C   BUSY  LEAVE      LNKS                   FREE LINK
    21         LOGIC R    *1                     ORIGIN HANGS UP
    22         TERMINATE  1                      BUSY CALLS
         *
    LNKS  STORAGE    10                     NO. OF LINKS
         *
     1         VARIABLE   X1*RN1/100C&1          PICK A LINE
     2         VARIABLE   X1-2*S$LNKS-2          COUNT NO. OF FREE LINES
         *
         INITIAL    X1,5C                  SET NO. OF LINES
         *
         START      10,NP                  INITIALIZE, SUPPRESS PRINT
         RESET                             WIPE OUT STATISTICS
         START      100C                   MAIN RUN
```

Figure 12-7. Coding of telephone system—model 1.

The following coding will place line numbers chosen at random from 50 lines in parameters numbers 1 and 2:

```
        ASSIGN      1,V1
        ASSIGN      2,V1
    1   VARIABLE    X1★RN1/1000+1
        INITIAL     X1,50
```

The same variable statement may be used for both assignments because each reference to the VARIABLE will produce a different random number.

With this method of generating the call origin and destination, it is possible that the origin of the call is already busy. A GATE block checks whether the selected origin is busy by using indirect addressing. If it is, the call is returned for reassignment. Should all lines be busy, this could cause an endless loop, so before assigning the origin, a check is made at a TEST block to ensure that at least two lines are not busy. The TEST block uses VARIABLE 2 to make the check and, if it finds the conditions unsatisfactory, the call is abandoned. It is also possible that the second ASSIGN block will choose the destination to be the same as the origin. This is checked at another TEST block which compares parameter 2 with parameter 1. If they are equal, the transaction is returned to the second ASSIGN block to reassign the destination.

When a valid call has been generated, the model makes the calling line busy by setting a switch, using indirect addressing. It then checks whether a link is available by attempting to enter the storage called LNKS whose capacity equals the number of links; in this case, 10. If the transaction cannot enter, the call is sent to a TERMINATE block called BLKD and the call is lost after the calling line switch is reset. Transactions that get a link check whether the called party is busy by using a GATE block, again using indirect addressing. If the line is busy, the call is lost and it goes to a location BUSY where the transaction terminates after resetting the calling line switch and returning the link. Otherwise, the call is established by setting the logic switch corresponding to the destination. An ADVANCE block represents the expenditure of time during the call, using function number 1 with a mean of 120. When the transaction leaves the ADVANCE block, the call is finished and the transaction proceeds to disconnect the call by resetting both logic switches, releasing the link, and terminating.

12-10
Set Operations

An important requirement in a simulation language is the capability of handling sets of temporary entities which have some common property. In GPSS, transactions that are blocked are automatically entered and removed from sets with a FIFO discipline. The program also has a way of allowing the user to control the sets so that more complex queuing disciplines can be simulated. A number of *user chains* are made available and a transaction is placed on a chain when it enters a LINK block. Field A carries the number (or name) of the chain and field B indicates the queuing discipline. The words FIFO or LIFO result in the disciplines they name. If Pn is used, the transactions are ordered by ascending values of parameter number n, with a FIFO rule for transactions having the same value. While on the chain, the transactions remain at the LINK block.

To correspond to the LINK block, there is an UNLINK block which allows a transaction (not on the chain) to remove transactions from the chain. The block names the chain in field A, and in field B gives the location to which unlinked transactions are to go. Field C says how many transactions are to be removed. The count can be an integer, the value of an SNA or the word ALL can be used to remove all the transactions. If only these first three fields are specified, the program removes transactions from the beginning of the chain. However, removal can be made to depend upon the value of any parameter of the transactions on the chain. Field D carries the number of the parameter to be examined and field E carries the value the parameter must have for the transaction to be removed. If field F is used, it provides a location to which the transaction entering the UNLINK block will go if it does not find a transaction on the chain that meets the conditions for removal. If field F is not used, the unlinking transaction goes to the next block, as it always does if it removes a transaction. _i.e, can't be handled_

As an example of how these blocks are used, suppose that, in the telephone system, blocked calls wait for a link to become free with the following service rules. Line 1 belongs to the company president. If there is an incoming call for line 1 and line 1 is free, the next free link goes to that call. Otherwise, the link goes to the call with the lowest origin number.

When transactions are blocked, they are now sent to a LINK block (see Fig. 12-8) which puts them on a chain called WAIT in ascending order of call origin (parameter 1). When a call terminates and releases a link, it checks to see if calls are waiting by going to a TEST block that compares an SNA called CHn against 0. This SNA is equal to the number of transactions on chain n. If there is a waiting call, the transaction goes to a GATE block to check whether line 1 is free. If so, it goes to the first UNLINK block of Fig. 12-8, which looks for transactions on the chain with the destination (parameter 2) equal to 1. If there is such a call waiting, it will be unlinked and sent to GETL for connection. If there is more than one call for line 1, the one that has waited longest will be chosen. When line 1 is not free, the unlinking transaction goes to the second UNLINK block which takes the first transaction on the chain; that is, the one with the lowest origin. If the transaction goes to unlink a call from line 1 and does not find one, it is sent to the second UNLINK block to unlink the call with the lowest origin. Figure 12-9 shows the coding for the problem in its new form. Calls meeting a busy condition are now sent to the TEST block at CKCH because they may have come off the chain. If so, another waiting call should be given a chance to use the link. Notice also that variable 2 has been changed, so that the TEST block, which checks whether it is feasible to generate a new call, now takes account of the waiting calls.

An alternative way of using the UNLINK block allows the unlinking to depend upon system conditions. This method makes use of *Boolean variable* statements. These are similar to variable statements but, instead of combining simple SNA's, they use the conditional test phrases of the GATE and TEST blocks. Each term of a Boolean variable is a single test that could be made by a GATE or TEST block. The terms are combined with the operators ★ and + for AND and inclusive OR, respectively. The individual terms take the value 1 or 0 according to whether the test is true or false. The values are combined by the

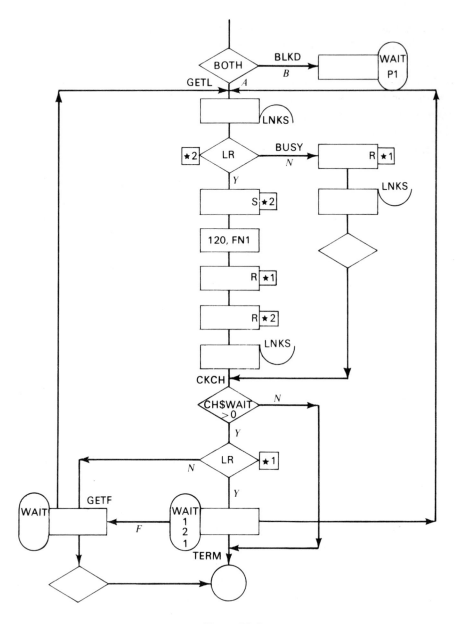

Figure 12-8

rules of the operators. For example, consider the following coding:

1   BVARIABLE   P2'E'1★LR1+P2'E'2★LR2

The Boolean variable will be equal to 1 (or true) if parameter number 2 equals 1 and logic switch number 1 is reset, or if parameter number 2 equals 2 and logic

```
BLOCK
NUMBER  *LOC    OPERATION  A,B,C,D,E,F,G              COMMENTS
        *
        *       SIMULATION OF TELEPHONE SYSTEM - MODEL 2
        *
        1       FUNCTION   RN1,C24              FUNCTION FOR I/A INTERVAL
       0.0,0.0/0.1,0.104/0.2,0.222/0.3,0.355/0.4,0.509/0.5,0.69
       0.6,0.915/0.7,1.2/0.75,1.38/0.8,1.6/0.84,1.83/0.88,2.12
       0.9,2.3/0.92,2.52/0.94,2.81/0.95,2.99/0.96,3.2/.97,3.5
       0.98,3.9/0.99,4.6/0.995,5.3/0.998,6.2/0.999,7/0.9997,8
        *
1               GENERATE   12,FN1               CREATE CALLS
2               TEST G     V2,2,ABND            TEST IF SYSTEM IS FULL
3       ASN1    ASSIGN     1,V1                 PICK ORIGIN
4               GATE LR    *1,ASN1              TEST FOR BUSY
5       ASN2    ASSIGN     2,V1                 PICK DESTINATION
6               TEST NE    P1,P2,ASN2           RETRY IF DEST = ORIGIN
7               LOGIC S    *1                   MAKE ORIGIN BUSY
8               TRANSFER   BOTH,,BLKD           TRY FOR LINK
9       GETL    ENTER      LNKS                 GET LINK
10              GATE LR    *2,BUSY              TEST FOR BUSY
11              LOGIC S    *2                   MAKE DEST. BUSY
12              ADVANCE    120,FN1              TALK
13              LOGIC R    *1                   ORIGIN HANGS UP
14              LOGIC R    *2                   DEST. HANGS UP
15              LEAVE      LNKS                 FREE LINK
16      CKCH    TEST G     CH$WAIT,0,TERM       TEST IF CALLS ARE WAITING
17              GATE LR    1,GETF               SEE IF LINE 1 IS FREE
18              UNLINK     WAIT,GETL,1,2,1,GETF CONNECT CALL TO 1
19      TERM    TERMINATE  1
20      GETF    UNLINK     WAIT,GETL,1          CONNECT FIRST WAITING CALL
21              TRANSFER   ,TERM
        *
22      ABND    TERMINATE                       ABANDON CALL
        *
23      BLKD    LINK       WAIT,P1              LINK IN ORDER OF CALL ORIGIN
        *
24      BUSY    LOGIC R    *1                   CALLER HANGS UP
25              LEAVE      LNKS                 FREE LINK
26              TRANSFER   ,CKCH                GO TO TEST FOR WAITNG CALLS
        *
        LNKS    STORAGE    10                   NO. OF LINKS
        *
        1       VARIABLE   X1*PN1/10000&1       PICK A LINE
        2       VARIABLE   X1-2*S$LNKS-CH$WAIT-2
        *
                INITIAL    X1,50
        *
                START      10,NP                INITIALIZE, SUPPRESS PRINT
                RESET                           WIPE OUT STATISTICS
                START      1000                 MAIN RUN
```

Figure 12-9. Coding of model 2.

switch number 2 is reset. Note that the TEST block conditions are placed between single quotes.

If the first UNLINK block of Fig. 12-8 has BV1 in field D (and nothing in field E), it will unlink transactions that meet the stated condition. With this form of the UNLINK block, it is not necessary to include the GATE block checking for line number 1 to be free.

Two other ways of organizing sets in GPSS will be briefly described, but the block types employed will not be described in detail. Transactions that are on a chain remain static until they are unlinked. It is sometimes necessary to identify members of a set that continue to move around the system. For example, it may be necessary to identify all the jobs in a factory for one customer, or all the cars of a given make. The GPSS program defines a number of *groups* for forming such sets. A block type JOIN allows a transaction to make itself a member of a group and then to proceed to the next block. Another block type REMOVE allows a transaction to remove itself from the group; it can be used to allow one transaction to remove others in much the same manner as transactions are unlinked from a chain. It is possible to SCAN the group for particular members, ALTER the parameters of the group members, or EXAMINE a transaction's group membership.

A common reason for wanting to form mobile transaction sets is that they represent inter-related tasks that must be coordinated. The making of a product, for example, may involve many operations, some of which can proceed independently; but some, such as an assembly, require that other operations be finished first. A block type called SPLIT allows one transaction to produce many copies which are automatically linked as a set but may proceed independently. A block type called ASSEMBLE will gather a given number of copies and merge them into one. It is also possible to synchronize the movement of copies with the use of a MATCH block, normally used in pairs. Copies arriving at a MATCH block must wait until a specified number of copies have arrived at another MATCH block before all the waiting copies can proceed.

Exercises
Draw GPSS block diagrams and write programs for the following problems:

12-1   Customers arrive at a single server counter with an average inter-arrival time of 20 ± 10 seconds. They purchase from 1 to 4 items with the following probabilities:

1	0.5
2	0.2
3	0.2
4	0.1

It takes 5 seconds to purchase each item. Tabulate the distribution of time for serving the first 100 customers.

12-2   Modify the supermarket problem of Fig. 12-2, so that 10% of the customers do not use a basket but otherwise are the same as other customers.

12-3   Jobs are passed into an office at the rate of one every 15 ± 5 minutes. Normally they go to clerk A, who takes 10 ± 3 minutes and then go to clerk B who takes 5 ± 2 minutes. However, if clerk B is busy at the time the job is brought to the office, the entire job is given to clerk C who takes 20 ± 10 minutes. Find out how many of the first 1,000 completed jobs will have been handled by clerk C. Assume each clerk handles one job at a time and assume that work can be stacked between clerks A and B.

12-4 People arrive at a bus stop at the rate of one every 20 ± 15 seconds. They queue for the bus unless the queue already has 10 people, in which case they walk away. A bus arrives every 5 ± 1 minutes. The people waiting board the bus one at a time, taking 5 seconds each. The bus waits for at least 20 seconds and it leaves as soon as the people stop boarding. Simulate 10 busloads of people. (Use separate transaction streams to represent people and buses and give a higher priority to the transactions representing the people. Use a logic switch to represent the presence or absence of a bus.)

12-5 Messages are being generated at the rate of one every 7 ± 3 seconds. They are to be transmitted to 1 of 4 destinations with the following probabilities:

1	0.2
2	0.3
3	0.35
4	0.15

All messages are first sent by a single main line to a switching center which then sends them on to their destinations by way of individual lines. Only one message can be on each of the lines at any time. If necessary, the switching center can store messages to await a free line. The messages are uniformly distributed in size from 10 to 100 characters. The main line can send messages at the rate of 10 characters a second while the other lines can only send them at the rate of 5 characters a second. Simulate the transmission of 1,000 messages and measure the queues on each line. (Use parameters and indirect addressing.)

12-6 In the telephone system model of Fig. 12-6, assume that 20% of the calls are long distance calls; the rest are local calls. Long distance calls are charged $1 each and local calls are charged 10¢ each. No charge is made for calls not connected. Modify the program to measure the telephone company's revenue.

12-7 A segment of a railroad consists of a single track section followed by a double track section. Only one train at a time can enter the single track section. As each train clears the single track section, a switch is thrown so that alternate trains go to each section of the double section. There is no limit on how many trains can be in the double train track section. Assume trains arrive every 10 ± 5 minutes and take 8 ± 4 minutes to clear the single track and 18 ± 9 to clear the double track section. Simulate the movement of 1,000 trains through the system and measure the queue that forms for the single track.

13

INTRODUCTION TO SIMSCRIPT

13-1

SIMSCRIPT Programs

SIMSCRIPT is a widely used simulation programming language designed principally for simulating discrete systems (1), (2), and (3). It is available on several different types of computer. As will be seen, it is a language similar in nature to FORTRAN. In fact, the earliest version of SIMSCRIPT operated by translating SIMSCRIPT statements into equivalent FORTRAN statements from which the program was compiled. This had the advantage of allowing most FORTRAN statements to be mixed with SIMSCRIPT statements. A later version of SIMSCRIPT, called SIMSCRIPT 1.5, (4), is compiled directly into machine language rather than through FORTRAN, thereby achieving greater program efficiency; however, the capability of including machine assembly language statements in the program has been substituted. SIM-SCRIPT statements are compatible with SIMSCRIPT 1.5, but, due to some simplifications in notation, not all SIMSCRIPT 1.5 statements will run under SIMSCRIPT. The programs described here are for SIMSCRIPT 1.5. Where there are differences between the notations of the programs, they will be noted, referring to the earlier version as SIMSCRIPT 1. Otherwise, references to SIMSCRIPT can be assumed to refer to both versions.

SIMSCRIPT
System
Concepts

The viewpoint to be taken by the SIMSCRIPT user corresponds essentially to that used in Sec. 1-1 to discuss the general nature of systems. That is to say, the system to be simulated is considered to consist of entities having attributes that interact with activities. The interaction causes events that change the state of the system. In describing the system, SIMSCRIPT uses the terms *entities* and *attributes*. For reasons of programming efficiency, it distinguishes between *temporary* and *permanent* entities and attributes. The former type represents entities that are created and destroyed during the execution of a simulation, while the latter represents those that remain during the run. A special emphasis is placed on the manner in which temporary entities form sets. The user can define *sets*, and facilities are provided for entering and removing entities into and from sets.

The user must define all the entities by giving a name to each entity and its attributes. In SIMSCRIPT 1.5, names can be any number of characters, provided the first six are unique and there are no imbedded blanks. In SIMSCRIPT 1, names can have up to five characters, but they must not end in F or contain the combination XX.

Activities are described by *event routines*, each of which is given a name and individually programmed as a separate closed routine. The execution of an event routine represents an event in the system that occurs at a particular point in time. A distinction is made between endogenous events, which result from actions within the system, and exogenous events, which arise from actions in the system environment. Correspondingly, the event routines are described as being either *endogenous* or *exogenous routines*.

In summary, the concepts used in SIMSCRIPT are

> Entities
>> Permanent
>> Temporary
> Sets
> Event routines
>> Endogenous
>> Exogenous

To show how the program is used, the telephone system that has previously been programmed in FORTRAN and GPSS, will be programmed in SIMSCRIPT. The system will first be considered a lost call system, in which calls that cannot be connected are abandoned. (See Sec. 8-4 for a description of the system.) The following chapter will extend the model. Telephone calls will be regarded as temporary entities; the telephone lines, which remain fixed in number, will be permanent entities. The origin and destination of the calls are temporary attributes. In addition to defining attributes that describe the nature of the entity, it may also be necessary to define attributes for the gathering of statistics. If, for example, the time spent in a queue is to be measured, an attribute of the call must be defined for recording the time of entry into the queue. The formation of a queue is an example of the use of sets. All calls waiting for a link to become free, for example, would form a set in SIMSCRIPT. Examples of programming sets will be given in the next chapter.

The arrival of new calls would normally be regarded as an exogenous event, and is programmed as such, if a specific list of calls is prepared and entered into the simulation. If the simulation uses the boot-strap method of generating calls, in which the arrival of one call generates the next, then all arrivals, except the first, will be programmed as an endogenous event. This procedure will be followed in the present example. A better example of exogenous events would be the programming of random breakdowns in telephone service.

When a call has arrived, its connection and disconnection are endogenous events and two endogenous routines will be needed for these purposes.

13-3
The Definition Form

A special *definition form* is filled out to describe the variables of the system. The names of the variables are recorded, along with their data formats. Variables may be floating-point or integer. It is possible to pack integers either one, two, or four to a word and floating-point variables one or two to a word. It is also necessary to indicate if a variable is to be signed or not.

All variables named on the definition form are called *system variables*. As such, they are available to all event routines. In addition, local variables can be used within individual event routines but they are only available to the routine in which they appear. The same name, however, can be used in different routines. Local variables cannot be packed and their mode is normally determined by their initial letter according to FORTRAN conventions.

Figure 13-1 shows the definition form completed for the telephone system. The names of the temporary entities and their attributes are entered in a section headed "Temporary System Variables." The first two lines of the section are for the call entity. The other entries in the section will be explained later.

The name ICALL will be used to mean a telephone call and it is entered in columns 4 to 8. Column 2 carries a T to indicate a temporary entity. The origin and destination of the call will be called IFROM and ITO, respectively, and they are entered in columns 20 to 24. They also have a T in column 18 to indicate that they are associated with a temporary entity. The number of words required for the entity is placed in column 9. A temporary entity must be described as 1, 2, 4, or 8 words. It is, in fact, possible to use more space by the use of satellite records but that feature will not be described here.

The location of each attribute within its temporary entity is given in column 26, while columns 27 through 29 give the packing. The notation 1/2, for

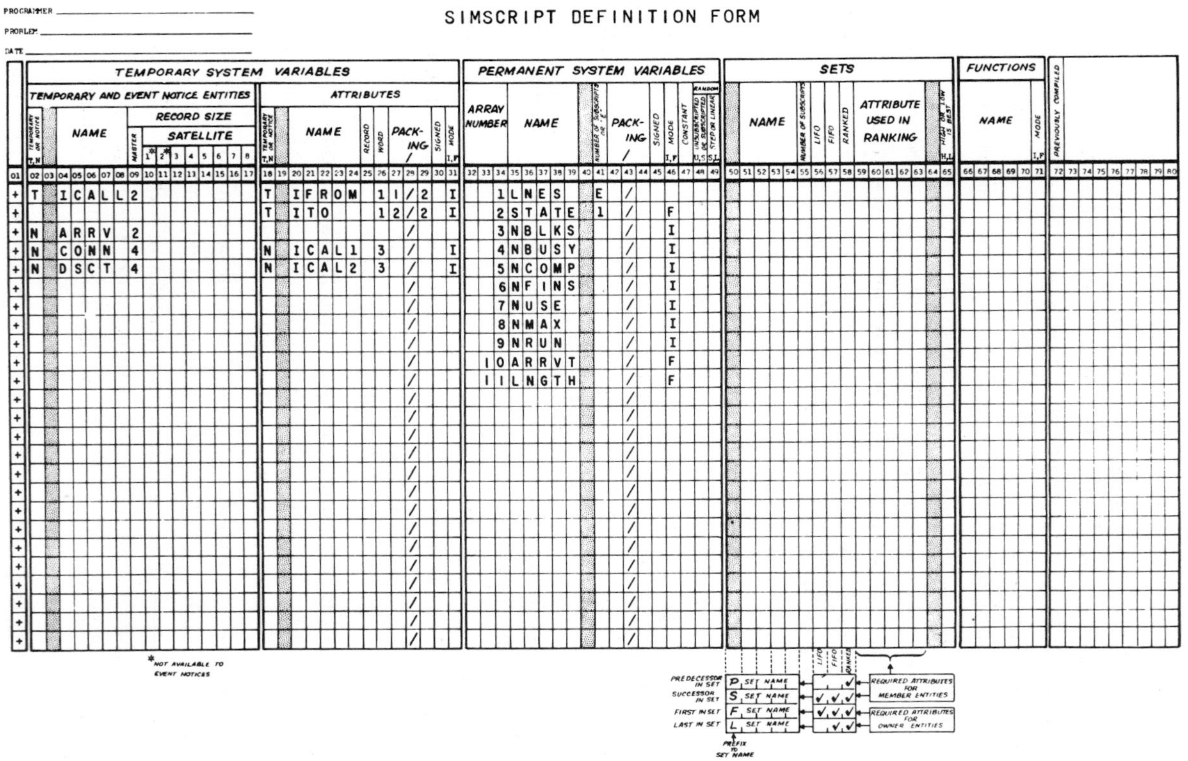

Figure 13-1. Definition form.

example, indicates that the attribute is the first half of the word. A blank in these columns means a full word. A 1 in column 30 means the attribute can be negative and an I or F in column 31 indicates integer or floating-point mode.

The names of permanent system variables go in columns 35 to 39. In this case, the first two entries are for a permanent entity LNES representing the lines, and its attribute STATE which records which lines are busy. The name of a permanent entity such as LNES is marked with an E in column 41. The purpose of this variable is to provide a name for the size of the arrays associated with the entity. The attribute STATE will be established as an array with one word for each line by putting a 1 in column 41. A word in STATE will be set to 1 if the line is busy and to 0 if it is not. Unlike FORTRAN, the size of the array does not have to be declared at the time the program is compiled; the size is supplied as part of the program initialization and associated with the name LNES. (It is also possible to have two-dimensional arrays which are associated with pairs of entities. For example, counts of the number of calls between the individual lines could be held in a two-dimensional array.)

Subscripted variables, such as STATE, may be packed. If required, packing is indicated in columns 42 to 44. The mode is given in column 46, and column 45 shows a 1 if the variable is to be signed.

The links will be represented by two numbers; the maximum number of links and the number currently in use. Locations for these attributes are defined as follows:

NMAX	Maximum number of links
NUSE	Number of links in use

A number of locations are needed as counters, so the following locations are defined:

NBLKS	Number of blocked calls
NBUSY	Number of busy calls
NCOMP	Number of completed calls
NFINS	Number of calls processed

The mean inter-arrival and service times must be read in; so also must the number of calls to be run. The locations for these numbers are

ARRVT	Mean inter-arrival time
LNGTH	Mean call length
NRUN	Number of calls to be run

Locations such as the counters, which are not associated with any particular entity, are called *system attributes*. The system attribute names are also entered in columns 35 through 39. They are single words that cannot be subscripted or packed, so columns 41 through 44 are blank. The mode and sign, however, can be selected in columns 45 and 46.

Because of their simple data structure, system attributes are processed more quickly than permanent entity attributes. For this reason, the links are represented as two system attributes rather than as a two-word array.

All permanent variables entered in columns 35 through 39 must be given a number in columns 32 through 34. The number is called an *array number* even if the variable is a single word. The purpose of the numbers is to link the definition form with the initialization form. The numbers do not have to be consecutive and there can be gaps in the number sequence. However, when initializing, all gaps must be accounted for by setting them to zero.

At initialization time, the value associated with LNES will be set to the number of lines and then used to establish the size of the array STATE that follows. For this reason, the permanent system entity name must carry a lower array number than that of any array attribute of the entity.

Other parts of the definition form will be explained later. When the form is completed, one card is punched for each line and the deck of cards forms the first part of the input deck. Note that the + sign in column 1 must be punched and the / in columns 28 and 43 should be punched if split words are used.

13-4

**Referencing
Variables**

Single word variables are, of course, referenced by using their names. Arrays such as STATE, are referenced with a name suffixed by an ordinal number. The state of the I th line is found with STATE(I), where I is an integer expression. The size of the array is associated with the name of the permanent entity to which the array belongs. The name of the permanent entity, however, must be preceded by an N. In this case, to find the number of lines, the variable NLNES is used.

Temporary entities are created and destroyed as the simulation proceeds. For each creation, the name of the entity type is specified and the program refers to the data structure given on the definition form. A block of words is created; and, since the number of blocks will fluctuate, they are kept in lists. Each member of the lists is identified by a pointer, and when it is being processed, a variable must be set aside to hold the pointer. The temporary entity name that appears on the definition form is a generic name for the type of entity. The variable holding the pointer is the name of a particular entity. As the entity moves around the system, the pointer location is changed so the entity name can change.

It is permissible, however, to have a local variable in each routine that has the same name as the generic name of the entity type. By the use of this feature,

Introduction to SIMSCRIPT / Ch. 13

the entity moving around the system can continue to be called by the generic name. Each routine in the telephone system simulation will have a local variable called ICALL so that the telephone call being processed can usually be called ICALL. A pointer will sometimes have to be stored in a location whose name is not the generic name; in particular, when the identity of the call is being passed from one routine to another. It is important in SIMSCRIPT programming to keep track of the name of a temporary entity as it proceeds through the system.

To refer to an attribute of a temporary entity, the name of the attribute is given followed by the name of the entity, i.e., the variable holding the pointer to the entity. The destination of a call, for example, can be referenced by IFROM(ICALL), assuming the pointer to the call is the variable ICALL. A powerful feature of SIMSCRIPT is that references can be indexed to any depth. Suppose, for example, the state of the line to which a call is going is required. The entry in the STATE table corresponding to the line number can be referenced by STATE(IFROM(ICALL)).

13-5
Organization of a SIMSCRIPT Program

Since the event routines are closed routines, some means must be provided for transferring control between them. The transfer is effected by the use of *event notices* which are created by the event routines when they determine that an event is scheduled. At all times, an event notice exists for every endogenous event scheduled to occur, either at the current clock time or in the future. Each event notice records the time the event is due to occur and the event routine that is to execute the event. If the event is to involve one of the temporary entities, of which there may be many copies, the event notice will usually identify which one is involved.

The general manner in which the simulation proceeds is illustrated in Fig. 13-2. The event notices are filed in chronological order. When all events that can be executed at a particular time have been processed, the clock is updated to the time of the next event notice and control is passed to the event routine identified by the notice. These actions are automatic and do not need to be programmed. Event notices do not usually go to more than one activity in the manner of a GPSS transaction. Having activated the routine, the event notice has served its purpose and is usually destroyed. Otherwise, expended event notices can eventually fill all available space.

If the event executed by a routine results in another event, either at the current clock time or in the future, the routine must create a new event notice and file it with the other notices. For example, in the telephone system, the connection of a call implies its disconnection at a later time, so the routine responsible for connecting the call will be responsible for producing the event notice that schedules the disconnection.

In the case of the exogenous events, a series of *exogenous event cards* are punched; one for each event. The cards are similar to event notices in that they give the time the event is to occur and identify the exogenous routine to execute the event. All exogenous event cards are sorted into chronological order and they are read by the program when the time for the event is due.

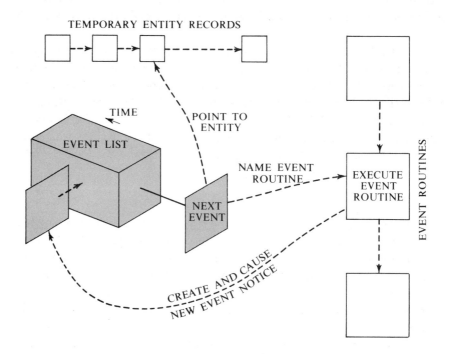

Figure 13-2. SIMSCRIPT execution cycle.

There must be at least one exogenous routine to introduce the first event into the system. Typically, this first event is used to start an endogenous routine which creates the first entity to be processed by the system and then continues to produce further entities in the bootstrap method described in Sec. 8-3. This procedure will be followed in the present problem.

An exogenous routine, to be called START, will be used to activate an endogenous routine called ARRV. The routine START will execute a single exogenous event which begins the simulation by transferring control to ARRV. Thereafter, the routine ARRV will produce a stream of calls from the inter-arrival statistics. The routine ARRV is therefore an endogenous routine. An endogenous routine called CONN will attempt to connect the call. When a connection had been made, another endogenous routine called DSCT will disconnect the call.

The event routines that need to be written are, therefore,

> Exogenous Routine
> > START   To begin the simulation
>
> Endogenous Routines
> > ARRV   To create calls
> > CONN   To connect calls
> > DSCT   To disconnect calls

Figure 13-3 illustrates the event routines that have been defined and shows the flow of event notices between the routines. The exogenous routine START is activated by a single exogenous event card.

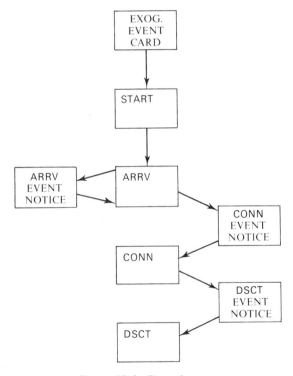

Figure 13-3. Flow of events.

One type of event notice is defined for each endogenous routine. From the point of view of programming, event notices are similar to temporary entities. They are described on the definition form in the same way as a temporary entity except that columns 2 and 18, which carry a T for a temporary entity, carry an

N for an event notice. There are, however, some important restrictions. Each type of event notice must have the *same name as the endogenous routine* it is to activate. In addition, an event notice must have at least two words and *the first two words must be left free* for use by the SIMSCRIPT program. Event notices can have attributes but, if they do, they must start in word three or higher.

A typical use of event notice attributes is to carry the identification of a temporary entity. The routine that schedules the event notice will store in the attribute the pointer to a particular temporary entity. The routine that executes the event can recover the identity of the particular entity to be processed by reading the attribute of the event notice. This is, in fact, the only use that will be made of event notice attributes in the present problem, but they can be used to pass any information from routine to routine. In this respect, the attributes of an event notice are similar to the parameters of a GPSS transaction. They can, for example, effect indirect addressing by identifying not only the temporary entity involved in an event but also the permanent entities.

For the present problem, three event notice types need to be defined; one named after each of the three endogenous event routines. The definitions are shown in Fig. 13-1 following the definition of the temporary entity ICALL. The event notice called ARRV is defined with the minimum size of two words. It is not necessary to assign attributes since the only purpose of the notice is to activate ARRV whenever a new call is to be produced. The other two event notices, CONN and DSCT, both need one attribute for identifying the call to be connected or disconnected. The attributes will be called ICAL1 and ICAL2 respectively. Note that the size of these notices is four words, since this is the next permissible size beyond the minimum of 2.

13-7

Programming Event Routines

Event routines are programmed with a series of statements written to the same format as FORTRAN statements. Column 1 of the coding form is left blank, unless a line of comment is to be introduced, in which case column 1 carries a C punch. The entire line will then be printed in the program listing.

The statements may be named or numbered in columns 2 to 6. (SIMSCRIPT 1 allows numbers only.) The numbers (or names) are local to the routine in which they appear. Column 6 is left blank unless a statement needs to be continued from a preceding card. Any mark in column 6 will indicate a continuation card. The statement is written between columns 7 to 72 inclusive. Any number of initial blanks are permitted in the statement field, allowing statements to be indented for clearer reading. In addition, one or more blanks is left between individual words of commands to make reading easier.

A separate subprogram is written for each event routine. Each routine is named and begins with one of the statements:

ENDOGENOUS EVENT "name"
EXOGENOUS EVENT "name"

Control is returned from the routine to the simulation algorithm by the command RETURN. This might occur at several places within the routine if the logic requires several different points of exit. The last statement of the routine must be END to indicate completion of the coding for the routine.

13-8

Management of Temporary Entities and Event Notices

Routines enter temporary entities into the simulation with a command called CREATE and remove them with a command called DESTROY. The most common way of using the CREATE command assumes that there is a local variable with the generic name of the type of entity it creates. In fact, the program will automatically create such a local variable if it has not been defined. In the telephone system, for example, a call can be created by the command

CREATE ICALL

The pointer to the created call will be placed in a local variable ICALL and the call can be referenced by that name.

If desired, the command can be extended to a long form:

CREATE ICALL CALLED SMITH

The name SMITH is a local variable that will receive the pointer to the call. The long form of the command would usually only be necessary if a routine creates a second call before finishing the processing of the first. The second call could be created by a second command such as

CREATE ICALL CALLED JONES

The two calls can then be distinguished by their names SMITH and JONES.

The DESTROY command, similarly, can have a short form; for example

DESTROY ICALL

which assumes the routine is using the generic name as a local variable. It can also have a long form, such as,

DESTROY ICALL CALLED SMITH

when the call has been given a specific name.

The same CREATE and DESTROY commands are used to enter and remove event notices. For these purposes, the commands may also take a short or long form. The long form would typically be used if the routine is to schedule simultaneously two or more events of the same type during one execution of the routine.

Another problem of identification that arises is that an event routine needs to recognize the event notice by which it was activated. Since an event notice must have the same name as the routine it activates, the program automatically creates a local variable with that common name in each routine. When the event notice activates the routine, its pointer is placed in this location. Consequently, an event notice is always recognized by its generic name within the routine it activates, even though the routine that scheduled the event notice may have used a different name. For example, when the routine CONN is activated, the pointer to the event notice will be in a local variable CONN. The identity of the call to be connected can be recovered from the event notice by referring to ICAL1(CONN).

The scheduling of an event is carried out with a CAUSE command which names the event notice type and the time at which the event is to occur. For example,

CAUSE DSCT AT TIME + LNGTH

will schedule a disconnection at time TIME + LNGTH. Upon execution of the command, the program automatically files the event notice with other event notices, in the manner illustrated in Fig. 13-2. At the appropriate time, the disconnection will become the next event and the event notice will be executed.

The example given assumes that the event notice is being called by its generic name. If it is not, the CAUSE command can take a long form, such as,

CAUSE DSCT CALLED SMITH AT TIME + LNGTH

TIME is an automatically defined floating-point system variable representing clock time. The time is normally expressed in units of days, hours, and minutes. The units can, however, be redefined. The time used in a CAUSE command can be an absolute time, expressed as a number or computed from an expression, or it can be expressed relative to current time, as has been done here.

It is not feasible to describe here all the commands of SIMSCRIPT. Table 13-1 gives a list of the commands that have been described so far, and others that will be described later (some in the next chapter). In the list, the term "variable" means the name of a variable, and "expression" means any group of variables combined by operators according to rules and conventions of FORTRAN. (An expression may, of course, be a simple variable.) Some of the other notations will be explained as the commands are described. For simplicity, only the long form of commands are given for those commands that have an option.

A command, LET, that has the following form:

$$\text{LET "variable"} = \text{"expression"}$$

will set the value of the named variable to the computed value of the expression. There is a similar command, STORE, which has the following form:

$$\text{STORE "expression" IN "variable"}$$

(In SIMSCRIPT 1, the expression can only be a simple variable.) The command appears to do the same thing as the LET command, although expressed in the reverse order. The two commands differ, however, in the way they respond to the mode of the data they transfer. The data components of the commands may differ in mode. If so, the LET command converts the data to the mode of the variable. The STORE command maintains the mode of the expression. This arrangement is made because of a common error in SIMSCRIPT programming. The user will frequently define a local variable in which to store the name of a temporary entity or event notice. Such a name is actually a pointer which must be an integer. Because the variable is local, its mode is implied by its first letter. The user may unwittingly define a floating-point variable and fail to notice the change of mode that occurs when using a LET command. The STORE command was designed to protect against this error, and the user is advised always to use the STORE command when transferring temporary event and event notice names. (The name is used as the expression of the command.)

There is an IF command that can compare any two expressions in a variety of ways. A typical use would be as follows:

IF "expression" "comparison" "expression", GO TO "statement no." (In SIMSCRIPT 1, the expressions must be enclosed in parentheses.) The comparisons that can be made are

GR	Greater than
GE	Greater than or equal to
EQ	Equal to
NE	Not equal to
LS	Less than
LE	Less than or equal to

CREATE "temporary entity or event notice" CALLED "variable"
DESTROY "temporary entity or event notice" CALLED "variable"
CAUSE "event notice" CALLED "variable" AT "floating-point time expression"
FILE "variable" IN "set"
REMOVE FIRST "variable" FROM "set"
REMOVE "variable" FROM "set"
LET "variable" = "expression", "any number of control phrases separated by commas"
STORE "expression" IN "variable", "any number of control phrases separated by commas"
FOR "local variable" = (expression)(expression)(expression)

$$\left.\begin{array}{l} \text{FOR EACH} \\ \text{FOR ALL} \\ \text{FOR EVERY} \end{array}\right\} \quad \text{"permanent entity" "local variable"}$$

$$\left.\begin{array}{l} \text{FOR EACH} \\ \text{FOR ALL} \\ \text{FOR EVERY} \end{array}\right\} \quad \text{"local variable"} \quad \left[\begin{array}{l} \text{OF} \\ \text{IN} \\ \text{ON} \\ \text{AT} \end{array}\right] \quad \text{"set"}$$

WITH "expression" "comparison" "expression"
OR "expression" "comparison" "expression"
AND "expression" "comparison" "expression"
DO
LOOP
IF "expression" "comparison" "expression"

$$\text{IF "set"} \left[\begin{array}{l} \text{IS} \\ \text{IS NOT} \end{array}\right] \text{EMPTY, "any statement"}$$

GO TO "statement number"

$$\text{FIND "variable"} \left[\begin{array}{l} \text{MAX} \\ \text{MIN} \end{array}\right] \text{OF "expression"}$$

WHERE "variable"
FIND FIRST, "one or more control phrases and WHERE phrase", IF NONE, "any statement"

$$\left.\begin{array}{l} \text{EXOGENOUS} \\ \text{EXOG} \end{array}\right\} \text{EVENT "event name"}$$

$$\left.\begin{array}{l} \text{ENDOGENOUS} \\ \text{ENDOG} \end{array}\right\} \text{EVENT "event name"}$$

RETURN
END
SUBROUTINE "subroutine name" ("arguments, if any")
CALL "subroutine name" ("arguments, if any")
REPORT "report name"
STOP

The statement appended to the IF command (after a comma) is executed if the tested condition is true; otherwise, control passes to the next command. The appended statement can be any of the SIMSCRIPT commands. The particular GO TO command used here executes an absolute transfer.

Figure 13-4 is a complete listing of the telephone system program, which will be explained in sections as the description of the simulation proceeds. The first part of the figure shows a listing of the cards that were punched from the definition form.

The first step in writing the program is to declare the number and names of all the routines. The section of Fig. 13-4 following the listing of the definition form shows how this is done for the telephone system. The number of both types of routine precede the statements EXOGENOUS and ENDOGENOUS. For the exogenous routines, an individual number is given in parentheses to each routine. The numbers need not be either sequential or in order. Note the END command at the end of this section.

The declaration of the routines is followed by the exogenous routine START, which will be activated at time zero by an exogenous event card. The card is described later in Sec. 13-12. The sole action of the START routine is to create an event notice ARRV and schedule it for immediate execution, and so create the first call.

The endogenous routine ARRV is shown next in Fig. 13-4. It creates a call and proceeds to select an origin and destination for the call. If the system is so full that only two lines or less are open, the call is abandoned. An origin is picked at random by using a SIMSCRIPT function RANDI(I,J) which produces an integral random number between I and J with equal probability. In this case, the range is 1 to NLNES, the number of lines. If the chosen line is already busy, another choice is made. The destination of the call is then picked at random. Should it happen to be the same as the origin, another choice is made.

When a legal call has been generated, the calling line is made busy. An event notice CONN is created and the identity of the call stored in the event notice attribute ICAL1. It is then scheduled to be executed immediately. Finally, another event notice, ARRV, is created to reactivate ARRV when the next arrival is due. A simple rectangular distribution of inter-arrival time is computed using a SIMSCRIPT function RANDM which gives a floating-point random number with uniform distribution between 0 and 1.

Note that the short form of the CREATE commands is being used both for creating the temporary entity ICALL and the event notice CONN. Consequently, the pointer to the call is a local variable called ICALL and the pointer to

```
+T ICALL2          T IFROM 11/2 I   1LNES   E
+                  T ITO   12/2 I   2STATE 1    F
+N ARRV 2                           3NBLKS      I
+N CONN 4          N ICAL1 3    I   4NBUSY      I
+N DSCT 4          N ICAL2 3    I   5NCOMP      I
+                                   6NFINS      I
+                                   7NUSE       I
+                                   8NMAX       I
+                                   9NRUN       I
+                                  10ARRVT      F
+                                  11LNGTH      F
C
C
        EVENTS
            1 EXOGENOUS
                   START(1)
            3 ENDOGENOUS
                   ARRV
                   CONN
                   DSCT
            END
C
        EXOG EVENT START
            CREATE ARRV
            CAUSE ARRV AT TIME
            RETURN
            END
C
        ENDOG EVENT ARRV
            DESTROY ARRV
            CREATE ICALL
            IF(NLNES-2*NUSE) GR 2,GO TO 1
            DESTROY ICALL
            GO TO 3
   1        LET IFROM(ICALL) = RANDI(1,NLNES)
            IF STATE(IFROM(ICALL)) GR 0.0,GO TO 1
   2        LET ITO(ICALL) = RANDI(1,NLNES)
            IF ITO(ICALL) EQ IFROM(ICALL),GO TO 2
            LET STATE(IFROM(ICALL)) = 1.0
            CREATE CONN
            STORE ICALL IN ICAL1(CONN)
            CAUSE CONN AT TIME
   3        CREATE ARRV
            CAUSE ARRV AT TIME + 2.0*ARRVT*RANDM
            RETURN
            END


        ENDOG EVENT CONN
            STORE ICAL1(CONN) IN ICALL
            DESTROY CONN
            IF NUSE EQ NMAX,GO TO 1
            IF STATE(ITO(ICALL)) GR 0.0,GO TO 2
            LET STATE(ITO(ICALL)) = 1.0
            LET NUSE = NUSE + 1
            CREATE DSCT
            STORE ICALL IN ICAL2(DSCT)
            CAUSE DSCT AT TIME + 2.0*LNGTH*RANDM
            RETURN
   1        LET NBLKS = NBLKS+1
            GO TO 3
   2        LET NBUSY = NBUSY+1
   3        LET NFINS = NFINS+1
            LET STATE(IFROM(ICALL)) = 0.0
            DESTROY ICALL
            IF NFINS LS NRUN,GO TO 4
            CALL FINAL
            STOP
   4        RETURN
            END
```

Figure 13-4. Telephone system in SIMSCRIPT.

```
      ENDOG EVENT DSCT
          STORE ICAL2(DSCT) IN ICALL
          DESTROY DSCT
          LET STATE(IFROM(ICALL)) = 0.0
          LET STATE(ITO(ICALL)) = 0.0
          LET NUSE = NUSE-1
          LET NCOMP = NCOMP+1
          LET NFINS = NFINS + 1
          DESTROY ICALL
          IF NFINS LS NRUN,GO TO 1
          CALL FINAL
          STOP
1         RETURN
          END


      REPORT FINAL
*                         SIMSCRIPT SIMULATION OF TELEPHONE SYSTEM
*                         NO. OF LINES                    ****
*                                                         NLNES
*                         NO. OF LINKS                    ***
*                                                         NMAX
*                         MEAN CALL INTERVAL              ***.*
*                                                         ARRVT
*                         MEAN CALL LENGTH                ***.*
*                                                         LNGTH
*                         NO. OF CALLS                    ****
*                                                         NRUN
*                         CLOCK TIME                      ****.*
*                                                         TIME
*                         CALLS COMPLETED                 ****
*                                                         NCOMP
*                         BUSY CALLS                      ****
*                                                         NBUSY
*                         BLOCKED CALLS                   ****
*                                                         NBLKS
                          END
                                                                    X
                                                            3       X
                                                                    X
                                                                    X
                                                                    X
                                                                    X
                                                                    X
                                                                    X
                                                                    X
                                                            3
                                                                    X
                                                            1
                                                                    X
                                                                    X
                                                                    X
                                                                    X
                                                                    X
                                                                    X
                                                                    X

      END
1     X     11
   1        R                               50
   2        1  Z    50   1
   3     7  0  Z
   8        0  R                            10
   9        0  R                            1000
  10        0  R                            12.0
  11        0  R                            120.0

   1     0  0  0
```

Figure 13-4. Telephone system in SIMSCRIPT (*cont.*).

the event notice is another local variable called CONN. The action of storing the identity of the call in the event notice is carried out by the following command which uses both these local variables:

STORE ICALL IN ICAL1(CONN)

The purpose is to carry to the connecting routine the identity of the call to be connected. The name of the call has now become ICAL1(CONN) and the call identity will be recovered in the CONN routine by using this name. (Within the ARRV routine, the call can continue to be called ICALL until a new call is created and the location ICALL is replaced with the pointer to the new call.)

The endogenous routine CONN, which connects a call is shown next in Fig. 13-4. Since the attempt at connection has been scheduled by the ARRV routine for immediate execution, these two routines could have been combined into a single routine. The routine CONN first extracts the identity of the call to be connected, by storing ICAL1(CONN) in ICALL. It then destroys the event notice. The routine checks if a link is available and if the called party is free. If so, the called party line is made busy. Note that it is not essential to use ICALL for storing the call identity. If the destruction of the event notice CONN is delayed, the called party line could be made busy by the command

LET STATE(IFROM(ICAL1(CONN))) = 1.0

An event notice DSCT is created, loaded with the identity of the call, and scheduled for execution at the end of the call. The call length is being represented here by a simple uniform distribution.

If the call cannot be connected, the appropriate counters are incremented and, if the required number of calls have been processed, the program is terminated by calling the routine FINAL and STOPing. The routine FINAL which prints the program output is described in the next section.

The routine DSCT for disconnecting the call is also shown in Fig. 13-4. It removes the call and increments the count of completed calls, which may result in the termination of the program. There are no further events for the call following its disconnection, so the call is destroyed and the routine does not issue an event notice.

13-11

Report Generation   A special form is filled out by the user to produce output reports. Fig. 13-5 shows the form filled out to produce an output report for the telephone system simulation. Each line of the report form is punched into a pair of cards. A line down the middle of the form separates the two cards. All left-hand cards are

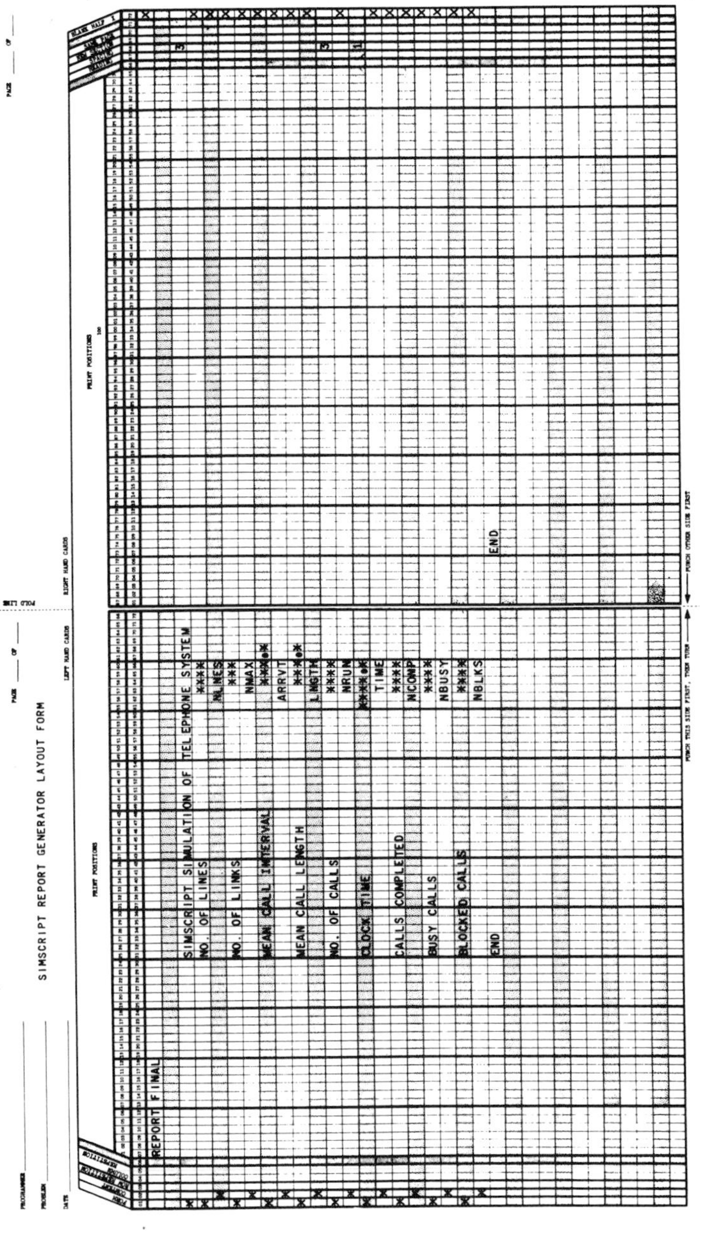

Figure 13-5. Report generation form.

punched first, and they are terminated with a card having the single word END. The right-hand cards follow, similarly terminated with an END card. The first left-hand card carries the word REPORT anywhere in the print position columns, followed by the report name. There can be many reports, each having a separate name, and they are executed by giving a CALL statement of the form

CALL   "report name"

Two types of line are filled out on the form. A form line, which has any mark in column 1, indicates the structure of the line of print. Text is placed exactly where it is to appear and fields of information are indicated by asterisks in the columns they are to occupy. When the number is decimal, the decimal point is indicated. The other type of line is a content line indicated by a mark in column 2. It immediately follows the form line with which it is associated. If the form line indicates only text, no content line is needed. Where there are fields of information, the content line gives the names of the fields in the order in which they are printed, separated by commas. It is not necessary, but it is convenient, to place the names in or near the columns in which the printed output is to appear.

If a line is so short that the right half is empty, this should be indicated by placing a mark in column 72 of the line. Various other controls, such as spacing lines and carrying headings from page to page, can be exercised by columns 67 to 70. A very convenient feature is that the printing of tables can be controlled by showing the content of a typical line and calling for row and column repetitions. The reader is referred to the manual for full details (1). Following the DSCT routine in Fig. 13-4 is a listing of the cards punched for FINAL.

Figure 13-7 shows an output produced with the report generator of Fig. 13-5.

13-12

Initialization   The initial values of all arrays must be established by completing an initialization form. Figure 13-6 shows the form completed for the telephone system simulation. The form is largely self-explanatory, although it allows for the setting up of tables of numbers, a feature that has not been described. All consecutive array numbers from 1 to the highest number used must be accounted for, even if they have not been used. Where an array is to be set to zero, the array number is put in columns 1–4 and a Z is put in column 13. Consecutively numbered arrays can, in fact, be set to zero with one card by showing the range of the array numbers in columns 1–8 and a Z in column 13. If a value other than zero is to be established, an R is punched in column 12, and the value is given in the field beginning in column 50. A single-subscripted permanent attribute (such as STATE), must have a 1 in column 10 and must show in columns 19 to 22 the array number of the permanent entity to which it belongs. In columns 15–18 is shown the number of rows to be given to the attribute. The last section

SIMSCRIPT INITIALIZATION FORM

PROGRAMMER ______
PROBLEM ______
DATE ______

SYSTEM SPECIFICATION CARD

(01–06)	MAXIMUM ARRAY NUMBER (07–12)	MINUTES PER HOUR (13–18)	HOURS PER DAY (19–24)	RANDOM ROOT (25–30)	INITIAL CONDITIONS TAPE (31–36)	EXOGENOUS EVENTS TAPE (37–42)	REPORT TAPE (43–48)	LINES PER PAGE (49–54)	COMMENT (55–72)	IDENTIFICATION (73–80)
1	11									
		60	24	1	5	5	6	55	← These values are automatically inserted if the field is left blank	

INITIALIZATION CARDS

ARRAY NUMBER FROM (01–04)	ARRAY NUMBER TO (05–08)	NUMBER OF SUBSCRIPTS (10)	READ IN VALUES R / SET TO ZERO Z (12–13)	ROWS: NUMBER OF ROWS (15–18)	ROWS: ARRAY NUMBER OF ATTRIBUTE EQUAL TO NUMBER OF ROWS (19–22)	COLUMNS: NUMBER OF COLUMNS (23–26)	COLUMNS: ARRAY NUMBER OF ATTRIBUTE EQUAL TO NUMBER OF COLUMNS (27–30)	LIST PACKING (32–34)	TABLE READ-IN: ACROSS ROWS R, DOWN COLUMNS C, NEW CARD AT BREAK N, FULL CARD F, PACKING 2,4, RAGGED TABLE R (36–41)	RANDOM LOOK-UP TABLES: UNSUBSCRIPTED U, SUBSCRIPTED S, STEP FUNCTION S, LINEAR INTERPOLATION L, INDIVIDUAL PROBABILITIES I, CUMULATIVE PROBABILITIES C (43–48)	INITIAL VALUE OR FORMAT FIELD DESCRIPTION (50–64)	COMMENT (65–72)	IDENTIFICATION (73–80)
1			R					/			50		
2		1	Z	50	1			/					
3	7	0	Z					/					
8		0	R					/			10		
9		0	R					/			1000		
10		0	R					/			12.0		
11		0	R					/			120.0		

Figure 13-6. Initialization form.

```
NO. OF LINES                        50
NO. OF LINKS                        10
MEAN CALL INTERVAL                  12
MEAN CALL LENGTH                   120
NO. OF CALLS                      1000


CLOCK TIME                     12095.7

CALLS COMPLETED                    707
BUSY CALLS                         246
BLOCKED CALLS                       47
```

Figure 13-7. Output of report generator.

of Fig. 13-4 shows a listing of the initialization cards punched from Fig. 13-6. The format for exogenous event cards is as follows:

Columns		
1– 3	Identification of exogenous event routine
4– 7	Day of occurrence
8–10	Hour of occurrence
11–12	Minute of occurrence
13–72	Additional data (if any)

The data in columns 13–72 can be broken into many different fields of various formats, and FORTRAN-like READ and FORMAT statements are used to control the data transfer. In this simulation, there will be only one card. It will be set to arrive at time zero and it carries no data.

13-13

Control Cards

A deck for the compilation and execution of a problem is prepared in the following way:

Definition cards
Declaration of the event routines
Event routines, subroutines, and reports (any order)
Data Card—Column 1 ★
        Column 7–72 DATA
System specification card
        Column 1 1
        Column 7–12 Maximum array number (right-justified)
Initial conditions deck
Blank card
Exogenous event card deck
End of file

SIMSCRIPT 1 (but not 1.5) must have a card, with the word SIMULATION in columns 7 to 72, included with the declaration of event routines.

13-1   Explain the changes that must be made to the telephone system model to introduce the following modifications. Identify clearly whether the changes involve entities, attributes, or activities. Treat each change separately.
(a) The program is to stop at clock time 50,000.
(b) Random breakdowns in service periodically cut off all calls in progress. Service is restored immediately.
(c) The accumulated time of all calls is to be recorded.
(d) Records are to be kept of all the individual link occupancies.

13-2   Change the telephone system simulation so that it takes 10 ± 5 time units to connect a call or find out that the called line is busy. A link is held during this period.

13-3   Change the telephone system simulation to charge calls at the rate of 1¢ per time unit and record the total cost of calls.

13-4   Change the telephone system simulation so that 10% of the calls are for weather or time information. These calls do not need a link and they can always be connected. Their length is 20 ± 10 time units.

13-5   Parts are produced by a machine tool at the rate of one every five minutes. Each part is inspected for 4 ± 3 minutes and 10% are rejected. Write a SIMSCRIPT program to simulate the system.

13-6   People arrive at a bus stop with inter-arrival times of 3 ± 1 minutes. A bus arrives with inter-arrival times of 15 ± 5 minutes. The bus has a capacity of 30 people and the number of seats occupied when the bus arrives is equally likely to be any number from 0 to 30. The bus takes on board as many passengers as it can seat and passengers that cannot be seated walk away. Simulate the arrival of 100 buses and count how many people do not get on board.

Bibliography

1   Markowitz, Harry M., Bernard Hausner, and Herbert W. Karr, *SIMSCRIPT—A Simulation Programming Language*, Englewood Cliffs, N. J.: Prentice-Hall, Inc., 1963.

2   Dimsdale, B., and H. M. Markowitz, "A Description of the SIMSCRIPT Language," *IBM Systems Journal*, III, No. 1 (1964), 57–67.

3   Markowitz, Harry M., "Simulating with SIMSCRIPT," *Management Science*, XII, No. 10 (June 1966), B396–405.

4   Karr, Herbert W., Henry Kleine, and Harry M. Markowitz, *SIMSCRIPT 1.5*, Santa Monica, Calif.: California Analysis Center, Inc., 1966.

14

MANAGEMENT OF SETS
IN SIMSCRIPT

SIMSCRIPT allows the formation of sets linking groups of temporary entities having a common property. The members of a set are maintained in order, according to one of three rules:

(a) First In First Out (FIFO)
(b) Last In First Out (LIFO)
(c) Ranked

The ranked rule allows any attribute of the set members to be selected as the basis for ordering the set; the member having the highest or lowest value of the attribute can be the first to be removed. Management of the sets is carried out with a series of commands that file and remove entities from the sets.

Sets can be defined with zero, one, or two subscripts. A zero subscript set is a single set considered as belonging to the system as a whole. A single-subscript set definition associates one set with every member of some permanent entity array, and a double-subscripted set definition associates one set with pairs of permanent entities. For example, in the telephone system, if blocked calls wait for a link to connect them, an unsubscripted set will be defined to hold all delayed calls. If calls finding the called party busy wait for the line to become free, separate queues must be formed for each line, and a single-subscripted set would be defined and called, say, QUE. There would be as many sets as lines, and reference to the queue for the Ith line would be made with the notation QUE(I). There can also be double-subscripted sets; for example, a set could be formed for all calls waiting to connect from line I to line J. The permanent entities with which the sets are associated are said to own the sets.

The members of the sets are organized in a list structure, as described in Sec. 10-6 and illustrated in Figs. 10-5 and 10-7. A pointer is used in each member of the set to indicate its successor. There must also be a header in which to indicate the first member of the set (or the fact that the set is empty). A ranked set and a FIFO set also need a trailer to indicate the last member. In the case of a ranked set, each member of the set must have a second pointer to indicate the predecessor member. The programmer must arrange space for these pointers and headers by assigning attributes for the pointers in the entities that belong to the set, and system attributes for the header and trailer.

Special rules apply to the naming of attributes used as pointers, headers, and trailers. According to its purpose, each set control attribute is formed by adding a single letter prefix to the name of the set. To define pointers in the members of the sets, the letter P or S is added for predecessor and successor pointers, respectively; for the header and trailer, the letters F and L are used to indicate first and last member, respectively. Since no variable can have a name with more than five characters, the rules for naming set control attributes require that a set cannot have a name with more than four characters. The rules that have just been described are summarized in a note printed at the bottom of the SIMSCRIPT definition form (see Fig. 13-1).

14-2
Telephone System Model 2

As an example of the way sets are used, the telephone system will be simulated for the case where calls that find all links busy wait for one to become available. We define a set WAIT in which to place the waiting calls. It is an unsubscripted set that belongs to the system. The calls will be served with a FIFO rule so two system attributes, FWAIT and LWAIT, are defined to hold pointers to the first and last in the set, and an attribute SWAIT is added to the temporary entity ICALL to carry a pointer to the successor in the set.

A listing of the program is given in Fig. 14-1. The changes that have been made in producing this model 2 from model 1 of the previous chapter will be described. The way the definition form was completed can be seen from the first part of the listing. The extra attributes mentioned above have been added. The name of the set, WAIT, was entered in columns 51 through 54 of the last line. It was marked as being a zero subscript set by a 0 in column 55, and a FIFO discipline was chosen by marking column 57. A counter NWAIT has been added, in which to keep a count of the number of calls that are waiting. The initialization now sets to zero the values of NWAIT, LWAIT, and FWAIT.

```
+T ICALL2        T IFROM 11/2 I   1LNES   E
+                T ITO   12/2 I   2STATE 1    F
+                T SWAIT 2    I
+N ARRV 2                         3NBLKS       I
+N CONN 4        N ICAL1 3    I   4NBUSY       I
+N DSCT 4        N ICAL2 3    I   5NCOMP       I
+                                 6NFINS       I
+                                 7NUSE        I
+                                 8NMAX        I
+                                 9NRUN        I
+                                1OARRVT       F
+                                1 1LNGTH      F
+                                12NWAIT       I
+                                13FWAIT       I
+                                14LWAIT       I         WAITO *
C
C
        EVENTS
             1 EXOGENOUS
                 START(1)
             3 ENDOGENOUS
                 ARRV
                 CONN
                 DSCT
             END
C
        EXOG EVENT START
             CREATE ARRV
             CAUSE ARRV AT TIME
             RETURN
             END
C
        ENDOG EVENT ARRV
             DESTROY ARRV
             CREATE ICALL
             IF(NLNES-2*NUSE-NWAIT) GR 2,GO TO 1
             DESTROY ICALL
             GO TO 3
        1    LET IFROM(ICALL) = RANDI(1,NLNES)
             IF STATE(IFROM(ICALL)) GR 0.0,GO TO 1
        2    LET ITO(ICALL) = RANDI(1,NLNES)
             IF ITO(ICALL) EQ IFROM(ICALL),GO TO 2
             LET STATE(IFROM(ICALL)) = 1.0
             CREATE CONN
             STORE ICALL IN ICAL1(CONN)
             CAUSE CONN AT TIME
        3    CREATE ARRV
             CAUSE ARRV AT TIME + 2.0*ARRVT*RANDM
             RETURN
             END


        ENDOG EVENT CONN
             STORE ICAL1(CONN) IN ICALL
             DESTROY CONN
             IF NUSE EQ NMAX,GO TO 1
        5    IF STATE(ITO(ICALL)) GR 0.0,GO TO 2
             LET STATE(ITO(ICALL)) = 1.0
             LET NUSE = NUSE + 1
             CREATE DSCT
             STORE ICALL IN ICAL2(DSCT)
             CAUSE DSCT AT TIME + 2.0*LNGTH*RANDM
             RETURN
        1    LET NWAIT = NWAIT+1
             FILE ICALL IN WAIT
             LET NBLKS = NBLKS+1
             GO TO 4
        2    LET NBUSY = NBUSY+1
             LET STATE(IFROM(ICALL))=0.0
             DESTROY ICALL
             IF WAIT IS EMPTY,GO TO 3
             REMOVE FIRST ICALL FROM WAIT
        LET NWAIT = NWAIT - 1
             GO TO 5
        3    LET NFINS = NFINS+1
             IF NFINS LS NRUN,GO TO 4
             CALL FINAL
             STOP
        4    RETURN
             END
```

Figure 14-1. Telephone system in SIMSCRIPT-model 2.

```
C
      ENDOG EVENT DSCT
          STORE ICAL2(DSCT) IN ICALL
          DESTROY DSCT
          LET STATE(IFROM(ICALL)) = 0.0
          LET STATE(ITO(ICALL)) = 0.0
          IF (NMAX-NUSE) GR 0,GO TO 2
          IF WAIT IS EMPTY,GO TO 2
          REMOVE FIRST IWAIT FROM WAIT
          LET NWAIT = NWAIT-1
          CREATE CONN
          STORE IWAIT IN ICAL1(CONN)
          CAUSE CONN AT TIME
2         LET NUSE = NUSE-1
          LET NCOMP = NCOMP+1
          LET NFINS = NFINS + 1
          DESTROY ICALL'
          IF NFINS LS NRUN,GO TO 1
          CALL FINAL
          STOP
1         RETURN
          END


      REPORT FINAL
*                              SIMSCRIPT SIMULATION OF TELEPHONE SYSTEM
*                              NO. OF LINES              ****
*  *                                                     NLNES
*                              NO. OF LINKS              ***
*  *                                                     NMAX
*                              MEAN CALL INTERVAL        ***.*
*  *                                                     ARRVT
*                              MEAN CALL LENGTH          ***.*
*  *                                                     LNGTH
*                              NO. OF CALLS              ****
*  *                                                     NRUN
*                              CLOCK TIME                ****.*
*  *                                                     TIME
*                              CALLS COMPLETED           ****
*  *                                                     NCOMP
*                              BUSY CALLS                ****
*  *                                                     NBUSY
*                              BLOCKED CALLS             ****
*  *                                                     NBLKS
                               END
```

```
                                                              X
                                                       3      X
                                                              X
                                                              X
                                                              X
                                                              X
                                                              X
                                                              X
                                                       3      X
                                                       1
                                                              X
                                                              X
                                                              X
                                                              X
                                                              X
                                                              X
                                                              X
      END
1     X    14
   1       R
   2    1  Z    50   1                          50
   3    7  J  Z
   8    0  R                                    10
   9    J  R                                    1JGC
  10    J  R                                    12.0
  11    0  R                                    120.0
  12   14  J  Z

   1    0  0  0
```

Figure 14-1. Telephone system in SIMSCRIPT-model 2 *(cont.)*.

The declaration of events and the routines START and ARRV remain unchanged, except that, in ARRV, the IF command that checks whether the system is too full to generate a call now takes account of the waiting calls, since they keep their originating lines busy.

If the routine CONN is able to connect a call, it proceeds as before. However, if a call is blocked, the routine now increments the counter NWAIT and files the call in the set WAIT. The file command has the form

FILE "variable" IN "set"

The variable is the name of the location holding the pointer of the entity to be filed.

The action taken when the program finds a busy line also changes. When the DSCT routine frees a link and finds there is a waiting call, it schedules the waiting call for immediate connection by CONN. If the call cannot be connected, service should be offered to the next waiting call, if any. After destroying a busy call, the CONN routine checks to see if there is a waiting call, with a command of the form

$$\text{IF "set"} \begin{bmatrix} \text{IS} \\ \text{IS NOT} \end{bmatrix} \text{EMPTY, "any statement"}$$

When there is a waiting call, the program goes to the next statement where it decrements the counter NWAIT and removes the first call with a command of the form

REMOVE FIRST "variable" FROM "set"

The variable is the location that receives the pointer of the removed entity. In this case, the variable is ICALL. The routine continues by transferring to statement 5 to attempt connection of the removed call. The process will continue until either a waiting call is connected or there are no more waiting calls.

When the routine DSCT disconnects a call, and therefore frees a link, it checks for waiting calls and removes the first one if there is one. The same REMOVE FIRST command is used as in CONN. Note, however, that the variable named in the command is IWAIT, so this becomes the name of the removed call. Although the entity being removed is a call, it cannot be given the generic name ICALL because the call being disconnected is still in existence and it has the name ICALL. In the CONN routine, the call being connected had the name ICALL but it was destroyed before removing a call from WAIT, so the name of the removed call can be ICALL. It is possible to rearrange the DSCT routine so that the disconnected call is destroyed before removing a waiting call. The program has been arranged in its present form, however, to illustrate the problem that arises when two entities of the same type are processed simultaneously.

Having removed a call, DSCT creates an event notice CONN, places the identity of the call in ICAL1, and schedules the call for immediate connection.

14-3

Searching Sets

It is often necessary to search a set for an entity with a particular combination of attribute values. It may not have been possible to order the entities of the set according to the values, so the queuing disciplines of the sets cannot be applied. Two commands are designed to see if a set contains an entity with particular values, and identify the entity. They are the FIND FIRST and FIND MAX (or MIN) commands. Both commands can also be used to search arrays of permanent entities, as will be explained in the next section. The commands operate with a number of control phrases that specify the nature of the search. The FIND FIRST command takes the form

> FIND FIRST, "one or more control phrases,"
> IF NONE, "any statement"

The control phrases name the set to be searched and the conditions being sought. If the set is empty or no member meets the conditions, the program executes the statement following the IF NONE phrase; otherwise, it proceeds to the next statement.

A FOR phrase, designed especially for use with sets, must be used. It has the form

$$\left.\begin{array}{c} \text{FOR EACH} \\ \text{FOR ALL} \\ \text{FOR EVERY} \end{array}\right] \text{``local variable''} \left[\begin{array}{c} \text{OF} \\ \text{IN} \\ \text{ON} \\ \text{AT} \end{array}\right] \text{``set''}$$

The local variable will be incremented as the search proceeds and will finally point to the entity that is found. It can be integer or floating-point, but it is advisable to keep it integer. As indicated, there are several choices of words; they are all equivalent in their actions. There is also a WHERE phrase, which takes the form

> WHERE "variable"

It may follow the FOR phrase. If the search is successful, the variable of the WHERE phrase is given the pointer to the entity that is found. It will then have the same value as the variable of the FIND FIRST command, but it will not change if the search is unsuccessful.

The search conditions are defined first by a WITH phrase, which has the form

WITH "expression" "comparison" "expression"

(In SIMSCRIPT 1 the expressions must be in parentheses.) The comparisons that can be made and the codes used are

GR	Greater than
GE	Greater than or equal
EQ	Equal to
NE	Not equal to
LS	Less than
LE	Less than or equal to

The WITH phrase follows a FOR phrase and is evaluated for each entity covered by the search. Suppose the set WAIT is to be searched for calls going to line number 5. The following statement could be used:

FIND FIRST, FOR EACH I OF WAIT, WITH ITO(I) EQ 5,
WHERE ICALL, IF NONE, GO TO 10

Note how the phrases are separated by commas.

The search will follow the order specified by the queuing discipline of the set, looking for calls going to line 5. It stops at the first such call and places a pointer to the call in the location ICALL. If there is no such call the program goes to statement number 10.

The conditions can be extended by using two other control phrases, AND and OR. They follow a WITH phrase and modify the conditions. They take the same form as a WITH phrase. Suppose the search is for a call not only going to line number 5 but also coming from line number 6 or 7. The following statement will perform the search:

FIND FIRST, FOR EACH I OF WAIT, WITH ITO(I) EQ 5,
AND IFROM(I) EQ 6, OR IFROM(I) EQ 7, WHERE ICALL,
IF NONE, GO TO 10

When an entity meeting the search conditions is found, it can be removed from the set by a REMOVE specific command:

REMOVE "variable" FROM "set"

The variable is the pointer to the entity to be removed.

The search of a set can be for an item with the maximum or minimum value of an expression. This is made with a command:

$$\text{FIND "variable"} = \begin{bmatrix} \text{MAX} \\ \text{MIN} \end{bmatrix} \text{"expression," "one or more control phrases," IF NONE, "any statement"}$$

There must be a FOR phrase and it can be modified by WITH, AND, and OR phrases. A WHERE phrase is usually included. The program will search all items specified by the FOR phrase and evaluate the expression for each item that meets the conditions laid down by the other control phrases. The maximum or minimum value found is placed in the variable of the command and, if a WHERE phrase is included, the location of the item that gave that value is placed in the variable of the WHERE phrase.

As an example, and also to demonstrate another queuing discipline, the telephone system will be reprogrammed with the assumption that waiting calls are served in the following way. Lines 1 and 2 belong to the company president and vice-president. A call waiting to be connected to either of them is offered the first link that becomes free provided they are not already busy. If there are calls waiting for both lines and they are both free, priority goes to the call for line 1. If there are no such calls, the link goes to the lowest numbered line that is attempting to *make* a call. (Compare with Sec. 12-10 which solves the same problem in GPSS.)

A listing of the program for this model 3 of the problem is given in Fig. 14-2, except that the routine FINAL, the initialization cards, and the exogenous event card are not given; they remain as in model 2. ARRV is also unchanged.

The set WAIT is now defined to be a ranked set by marking column 58 rather than 57. The waiting calls will be ranked according to the line number originating the call, with the lowest numbered line being given preference. The attribute IFROM is named in columns 59 to 63 and an L is placed in column 65. Note particularly that, although the definition form implies that column 64 is blank, it must contain an I or F to indicate the mode of the variable by which the entities are ranked. In this case it is an I.

In the routine CONN, the instruction that files waiting calls remains unchanged. There is sufficient information on the definition form for the program to interpret the request correctly. The actions of locating the correct place for the new call and adjusting the pointers are carried out automatically.

When a call has been found busy, and it has been determined that there are waiting calls, the program now executes a FIND MIN command which is modified by five phrases. It searches the set WAIT looking for calls that are directed to line 1 or line 2 and when it finds one, checks whether the line being called is free. When there are calls to both 1 and 2 that meet these conditions,

```
+T ICALL4         T IFROM 11/2 I     1LNES  E
+                 T ITO   12/2 I     2STATE 1   F
+                 T SWAIT  2   I
+                 T PWAIT  3   I
+N ARRV 2                           3NBLKS     I
+N CONN 4         N ICAL1 3   I     4NBUSY     I
+N DSCT 4         N ICAL2 3   I     5NCOMP     I
+                                   6NFINS     I
+                                   7NUSE      I
+                                   8NMAX      I
+                                   9NRUN      I
+                                  10ARRVT     F
+                                  11LNGTH     F
+                                  12NWAIT     I
+                                  13FWAIT     I
+                                  14LWAIT     I
+                                             WAITO  *IFROMIL
C
C
        EVENTS
            1 EXOGENOUS
                 START(1)
            3 ENDOGENOUS
                 ARRV
                 CONN
                 DSCT
        END
C
        EXOG EVENT START
            CREATE ARRV
            CAUSE ARRV AT TIME
            RETURN
        END
C
        ENDOG EVENT ARRV
            DESTROY ARRV
            CREATE ICALL
            IF(NLNES-2*NUSE-NWAIT) GR 2,GO TO 1
            DESTROY ICALL
            GO TO 3
1           LET IFRCM(ICALL) = RANDI(1,NLNES)
            IF STATE(IFROM(ICALL)) GR 0.0,GO TO 1
2           LET ITO(ICALL) = RANDI(1,NLNES)
            IF ITO(ICALL) EQ IFROM(ICALL),GO TO 2
            LET STATE(IFROM(ICALL)) = 1.0
            CREATE CONN
            STORE ICALL IN ICAL1(CONN)
            CAUSE CONN AT TIME
3           CREATE ARRV
            CAUSE ARRV AT TIME + 2.0*ARRVT*RANDM
            RETURN
        END


        ENDOG EVENT CONN
            STORE ICAL1(CONN) IN ICALL
            DESTROY CONN
            IF NUSE EQ NMAX,GO TO 1
5           IF STATE(ITO(ICALL)) GR 0.0,GO TO 2
            LET STATE(ITO(ICALL)) = 1.0
            LET NUSE = NUSE + 1
            CREATE DSCT
            STORE ICALL IN ICAL2(DSCT)
            CAUSE DSCT AT TIME + 2.0*LNGTH*RANDM
            RETURN
1           LET NWAIT = NWAIT+1
            FILE ICALL IN WAIT
            LET NBLKS = NBLKS+1
            GO TO 4
2           LET NBUSY = NBUSY+1
            LET STATE(IFROM(ICALL))=0.0
            DESTROY ICALL
            IF WAIT IS EMPTY,GO TO 3
            FIND IFIND = MIN OF ITO(I),FOR EACH I OF WAIT,WITH ITO(I) LE 2
1           ,AND STATE(ITO(I)) EQ C.O, WHERE ICALL IS NEXT,IF NONE,GO TO 3
            REMOVE ICALL FRCM WAIT
            LET NWAIT = NWAIT - 1
            GO TO 5
3           LET NFINS = NFINS+1
            IF NFINS LS NRUN,GO TO 4
            CALL FINAL
            STOP
4           RETURN
        END
```

Figure 14-2. Telephone system in SIMSCRIPT-model 3.

```
C
        ENDOG EVENT DSCT
            STORE ICAL2(DSCT) IN ICALL
            DESTROY DSCT
            LET STATE(IFROM(ICALL)) = 0.0
            LET STATE(ITO(ICALL)) = 0.0
            IF (NMAX-NUSE) GR 0,GO TO 2
            IF WAIT IS EMPTY,GO TO 2
            FIND IFIND = MIN OF ITO(I),FOR EACH I OF WAIT,WITH ITO(I) LE 2
     1      ,AND STATE(ITO(I)) EQ 0.0, WHERE IWAIT IS NEXT,IF NONE,GO TO 5
            REMOVE IWAIT FROM WAIT
            GO TO 6
     5      REMOVE FIRST IWAIT FROM WAIT
     6      LET NWAIT = NWAIT-1
            CREATE CONN
            STORE IWAIT IN ICAL1(CONN)
            CAUSE CONN AT TIME
     2      LET NUSE = NUSE-1
            LET NCOMP = NCOMP+1
            LET NFINS = NFINS + 1
            DESTROY ICALL
            IF NFINS LS NRUN,GO TO 1
            CALL FINAL
            STOP
     1      RETURN
            END

C
1      X    14
   1         R                                            50
   2      1  Z    50    1
   3      7 0  Z
   8      0 R                                             10
   9      0 R                                             1000
  10      0 R                                             20.0
  11      0 R                                             120.0
  12     14 0  Z

   1     0  0  0
```

Figure 14-2. Telephone system in SIMSCRIPT-model 3 (*cont.*).

it picks one to line 1. If the program is successful in finding a call that meets the conditions, it puts the pointer to the call in ICALL. It also puts the number of the called line in IFIND, although the information is not being used in this example. When a call is found, the program proceeds to the next statement, which is a REMOVE specific command.

The two statements in the DSCT routine of model 2 that search the set and remove a call are changed in the same way. The variable named in the WHERE phrase and the REMOVE specific command is IWAIT because the name ICALL is still assigned to the call being disconnected at the time a call is removed from the set.

If no call is found, the program goes to a REMOVE FIRST command. Since the calls have been filed in descending order of calling number, the command will take the call coming from the lowest numbered line. Whichever type of call is removed, the program proceeds as before by decrementing the count of waiting calls and transferring to statement 5 to attempt a connection.

14-4

Searching Arrays

Arrays associated with a permanent entity can be searched by a FIND FIRST or FIND MAX(MIN) command in the same way a set is searched, except that a different form of FOR phrase must be used. The entity form of the FOR phrase is

$$\left.\begin{array}{l}\text{FOR EACH}\\\text{FOR EVERY}\\\text{FOR ALL}\end{array}\right]\quad \text{"permanent entity"} \quad \text{"local variable"}$$

(In SIMSCRIPT 1, the variable must be integer.) The permanent entity whose array is to be searched is named so the program knows the size of the array. The local integer variable will be incremented during the search and, if an entry is found, it will contain the number of the entry. The control phrases WITH, AND, OR, and WHERE are used in conjunction with the FOR entity phrase in the same way as they are used when searching sets.

As an example, suppose the telephone model keeps records of how many calls have been made to each line and the program is to find the line that has received the most calls. An array called ICNT is defined as an attribute of LNES to hold the count of calls made to each line. The following command will search ICNT, place the number of the line with the most calls in IMOST and the number of calls to that line in ITTL:

FIND ITTL = MAX OF ICNT(I), FOR EACH LNES I, WHERE IMOST

14-5
DO Loops

A command DO can be used to control the execution of a string of statements. The end of the string is marked by a LOOP command. (In SIMSCRIPT 1, the command is DO TO, followed by a statement number assigned to the LOOP command.) Another form of FOR phrase is used to control the loop. It takes the form:

FOR "local variable" = (expression) (expression) (expression)

The modes may be integer or floating-point in SIMSCRIPT 1.5, but they must all be the same. (In SIMSCRIPT 1, they must be integer.)

For example,

FOR I = (1) (ITAB − 5) (2)

will execute the loop for I going from 1 to ITAB − 5 in increments of 2. If the increments are 1, the last expression can be dropped.

To execute a DO loop over an array or a set, the appropriate form of FOR phrase must be used, and it may be modified by the WITH, AND, and OR phrases. (In SIMSCRIPT 1 a DO TO loop over a set is terminated by a REPEAT command.)

If the actions to be performed in the DO loop can be performed by LET or STORE commands only, it is not necessary to establish the loop because both these commands can be controlled directly by any of the forms of the FOR phrases, modified by the other control phrases. For example, assume the array ICNT, holding the count of calls made to each line, is to be searched, and the number of lines that have received more than 10 calls is to be placed in NTEN. The following LET command, modified by an entity form of FOR phrase, can be used:

LET NTEN = NTEN + 1, FOR EACH LNES I, WITH ICNT(I) GR 10

The use of a set form of FOR phrase to control the execution of a command over a set is permitted. For example, suppose the set WAIT is to be searched to find the number of calls waiting for line 5. The following statement will place the count in NWT:

LET NWT = NWT +1, FOR EACH I OF WAIT, WITH ITO(I) EQ 5

14-6

Statistical Commands

Section 10-3 discussed how statistics on utilization and occupancy are derived. They involve keeping a count of how many items are in use at any time, keeping a record of the last time a change occurred, and, as each change is made, accumulating a quantity. (See Figs. 10-1 and 10-2.) A command called AC-CUMULATE carries out these actions in SIMSCRIPT. It names the variable to be monitored, a location for the accumulation, and a location for storing the last change time. An optional phrase will ADD to a counter. All variables to be named are floating-point. The following example is taken from the manual:

ACCUMULATE IDLE(MG), INTO CIDLE(MG),
SINCE LASTM(MG), ADD −1.0

This multiplies the value of IDLE(MG) by the time that has elapsed since LASTM(MG) and adds the result to CIDLE(MG). It then adds − 1 to IDLE(MG) and updates LASTM(MG) to the clock time. A list of totals can be accumulated in one command by placing lists of variable names in each of the places where variables are specified.

Another command, COMPUTE, will calculate a mean value, standard deviation, and variance of a set of numbers X(I). It takes the following form:

COMPUTE MX, SX, VX = MEAN, STD-DEV, VARIANCE
OF X(I), FOR EACH BASE I

The variables on the left of the equality sign name the locations to be used for the results. MEAN OF X(I), STD-DEV OF X(I), and VARIANCE OF X(I) appear if all three quantities are wanted. Any of them can be dropped. The variables on the left are matched with these key words. The command can be modified by control phrases in the same way as a LET command.

In exercises 14-1 through 14-4, modify the telephone system simulation to introduce the following changes:

14-1   Compute the average waiting time of all calls that have to wait, whether they are connected or not.

14-2   Compute the average and the maximum queue length.

14-3   Assign the length of the call as an attribute and organize the queue so that the call that will finish first is connected first.

14-4   Assume calls to line number 1 wait if the line is busy, keeping a link engaged. Organize the queue of calls for line 1 on a FIFO discipline.

14-5   Program the manufacturing shop model 2 of Fig. 11-5 in SIMSCRIPT.

14-6   Reprogram the bus service problem of exercise 13-6 assuming that passengers are to queue with a FIFO discipline and that they wait for the next bus if they are not able to board a bus.

15

VERIFICATION OF
SIMULATION RESULTS

With the introduction of stochastic variables into a simulation, the variables used to measure the system performance become random variables. The problem of gauging the significance of the results must be considered. The values measured are no more than a sample, and they must be used to estimate the parameters of the distribution from which they are drawn.

A simulation study is usually planned as a series of runs aimed at comparing a number of alternative systems or operating conditions. To help clarify the discussion, the term experiment will be used to mean the testing of one particular system operating under one set of conditions. The term simulation run will mean a single execution of one experimental configuration. An observation will be a single measurement of a system variable. An experiment is the collection of all runs with one system configuration and the study is the collection of all experiments.

The statistical problems associated with a simulation study have been conveniently classified by Conway (1) into two broad classes:

(a) Strategic planning problems, concerned with the design of a set of experiments.
(b) Tactical planning problems concerned with specifying how each experiment is to be conducted.

The strategic planning must determine the measures by which the system is judged and how to test the significance of differences in these measures. The tactical planning must decide how the measures are to be taken from each run and how many runs should be made for each experiment.

The strategic planning problem will not be discussed here since this is a problem that is largely independent of simulation techniques. A considerable body of literature exists on the design of experiments. In particular, Burdick and Naylor (2) have reviewed the available methods in the light of simulation studies and provided a substantial set of references to available works. Few methods have been specifically developed for application to simulation experiments, and many of the classical methods are not particularly well suited for this purpose because they assume the existence of independent data on which to base statistical tests. As will be seen, simulation data frequently do not meet this condition. However, as discussed by Conway (1) and Naylor (3), some of the more recently developed methods, such as sequential analysis and ranking methods, show promise for simulation applications.

15-2
Estimation Methods

Before discussing the tactical planning problem in the context of simulation runs, we review some of the statistical methods commonly used to estimate parameters from observations on random variables. Normally, a random variable x_i is drawn from a population that has a stationary probability distribution with a finite mean μ and variance σ^2. If n independent observations are made of the variable, the sample mean $\bar{x} = \dfrac{1}{n} \sum_{i=1}^{n} x_i$ is also a random variable. The central limit theorem establishes that the distribution of $\bar{x}$ tends to a normal distribution with mean μ and variance σ^2/n. It follows that the quantity

$$z = \frac{\bar{x} - \mu}{\sigma/\sqrt{n}}$$

is approximately normally distributed with a mean of 0 and a variance of 1. This distribution is the standardized normal distribution which was discussed in Sec. 7-9. The integral of the distribution from $-\infty$ to a value u is the probability that z is less than or equal to u. The integral is usually denoted by $\Phi(u)$ and tables of its value are widely available. Suppose the value of u is chosen so that $\Phi(u) = 1 - \alpha/2$, where $\alpha/2$ is some constant less than 1, and denote this value of u by $u_{\alpha/2}$. The probability that z is greater than $u_{\alpha/2}$ is then $\alpha/2$. The normal distribution is symmetric about its mean, so the probability that z is less than $-u_{\alpha/2}$ is also $\alpha/2$. Consequently, the probability that z lies between $-u_{\alpha/2}$ and $u_{\alpha/2}$ is $1 - \alpha$. That is,

$$\text{Prob}\left\{-u_{\alpha/2} \leqslant z \leqslant u_{\alpha/2}\right\} = 1 - \alpha$$

In terms of the sample mean, this probability statement can be written

$$\text{Prob}\left\{ \bar{x} - \frac{\sigma}{\sqrt{n}} u_{\alpha/2} \leqslant \mu \leqslant \bar{x} + \frac{\sigma}{\sqrt{n}} u_{\alpha/2} \right\} = 1 - \alpha$$

The constant $1 - \alpha$ (usually expressed as a percentage) is the confidence level and the interval

$$\bar{x} \pm \frac{\sigma}{\sqrt{n}} u_{\alpha/2}$$

is the confidence interval. The size of the confidence interval depends upon the confidence level chosen. Typically, the confidence level might be 90%, in which case $u_{\alpha/2}$ is 1.65. The statement then says that μ will be within the confidence interval $\bar{x} \pm 1.65\sigma/\sqrt{n}$ with probability 0.9; meaning that, if the experiment is repeated many times, the mean can be expected to fall within the confidence interval on 90% of the repetitions.

In practice, the population variance σ^2 is not usually known; in which case, it is replaced by an estimate calculated from the formula

$$s^2 = \frac{1}{n-1} \sum_{i=1}^{n} (x_i - \bar{x})^2 \tag{15-1}$$

The normalized random variable z based on the estimate of σ^2 is no longer distributed according to the standardized normal distribution; instead it has a Student-t distribution. The quantity $u_{\alpha/2}$ used in the definition of a confidence interval should then be derived from the integral of the Student-t distribution. The deviation between the two distributions diminishes as n increases, and, for sufficiently large n ($\geqslant 30$), the normal distribution can be used.

Expressed in terms of the estimated value of the population variance, the confidence interval for μ is

$$\bar{x} \pm \frac{s}{\sqrt{n}} u_{\alpha/2}$$

where $u_{\alpha/2}$ is based on a Student-t distribution when n is small, and on the normal distribution when n is large.

15-3

Simulation Run Statistics

The method of establishing a confidence interval outlined in the previous section was based upon two assumptions. It assumed that the distribution from which the observations were drawn was stationary, and it assumed that the observations were independent. Unfortunately, many statistics of interest in a simulation do not meet these conditions. To illustrate the problems that arise in measuring statistics from simulation runs, a specific example will be discussed.

Consider a single server system in which the arrivals occur with a Poisson distribution and the service time has an exponential distribution. The queuing discipline is first-in, first-out with no priority. Suppose the study objective is to measure the mean waiting time, defined as the time entities spend waiting to receive service and excluding the service time itself. The problem can be solved analytically. The solution was given in Eq. (7-8) and the probability of the waiting time was also plotted in Fig. 7-6.

In a simulation run, the simplest approach is to estimate the mean waiting time by accumulating the waiting time of n successive entities and dividing by n. This measure will be called the sample mean and denoted by $\bar{x}(n)$ to emphasize the fact that its value depends upon the number of observations taken. If x_i ($i = 1, 2, \ldots, n$) are the individual waiting times (including the value 0 for those entities that do not have to wait), then

$$\bar{x}(n) = \frac{1}{n} \sum_{i=1}^{n} x_i \qquad (15\text{-}2)$$

Waiting times measured this way are not independent. Whenever a queue forms, the waiting time of each entity on the queue clearly depends upon the waiting times of its predecessors. Any series of data that has this property of having one value affect other values is said to be *autocorrelated*. The degree to which the data is autocorrelated can be measured in ways that will be briefly described in a later section. In the particular problem being discussed, the auto-correlation increases rapidly as the utilization of the service facility increases.

Under broad conditions that can normally be expected to hold in a simulation run, the sample mean of autocorrelated data can be shown to approximate a normal distribution as the sample size increases (4). The usual formula for estimating the mean value of the distribution, Eq. (15-2), remains a satisfactory estimate for the mean of autocorrelated data. However, the variance of the autocorrelated data is not related to the population variance by the simple expression σ^2/n, as occurs for independent data. A term must be added to account for the autocorrelation. The term is positive in situations that normally occur in a simulation experiment so that, if it is ignored, the variance is under-estimated and an excessively optimistic confidence interval will be calculated.

Another problem that must be faced is that the distributions may not be stationary. In particular, a simulation run is started with the system in some initial state, frequently the idle state, in which no service is being given and no entities are waiting. The early arrivals then have a more than normal probability of obtaining service quickly, so a sample mean that includes the early arrivals will be biased. As the length of the simulation run is extended and the sample size increases, the effect of the bias will die out. For a given sample size starting from a given initial condition, the sample mean distribution is stationary; but, if the distributions could be compared for different sample sizes, the distributions would be slightly different. The analytical solutions previously quoted are for the steady state values to which the distributions converge as the sample size increases.

To illustrate these problems, Fig. 15-1 shows the results of measuring the sample mean waiting time for the single server system. Results of three runs are shown, each for the case where the system utilization is 0.5. For each run, the system was started in the idle state; the runs differ only in the fact that

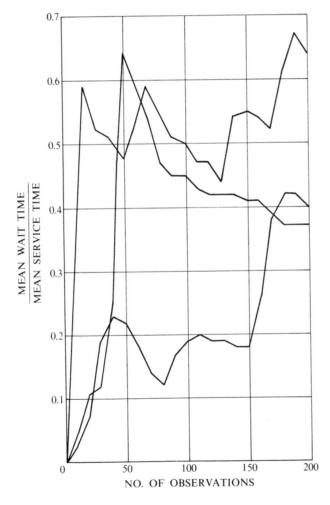

Figure 15-1. Variability of simulation results.

different random numbers were used. Observations were made as each tenth entity was served. The ratio of the mean waiting time to the service time is shown; in this case, the ratio should be 0.5. The variability of the sample mean is immediately apparent and the initial bias caused by starting the system from the idle state can also be seen.

The sample mean will ultimately settle to a steady value because it is a cumulative statistic. Figure 15-2, for example, shows the results for the same experiment conducted for 10,000 entities with samples at every 500. Even here, however, significant fluctuations can be seen after many entities have been measured. The fact that the accumulated sample mean tends to a steady value does not, of course, mean the waiting time tends to a steady value. The individual waiting times will continue to show their inherent variability no matter how many are measured, but, with the accumulation of observations, the variations of the sample mean will balance out.

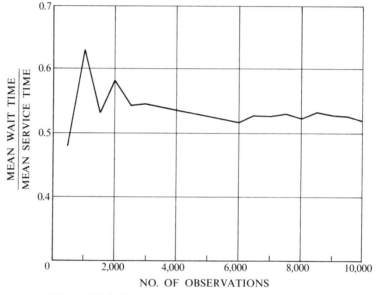

Figure 15-2. Stabilization of accumulated mean wait time.

15-4

Repetition of Runs

Figure 15-1 suggests one way of deriving a measure of the sample mean variance. Repeating the experiment with different random numbers for the same sample size n gives a set of independent determinations of the sample mean $\bar{x}(n)$. Even though the distribution of the sample mean depends upon the degree of autocorrelation, these independent determinations of the sample mean can properly be used to estimate the variance of the distribution. Suppose the experiment is repeated p times with independent random number series. Let x_{ij} be the ith observation in the jth run, and let the value of the sample mean for the jth run be $\bar{x}_j(n)$. Then, the estimate of the mean waiting time and its variance are

$$m(n) = \frac{1}{p} \sum_{j=1}^{p} \bar{x}_j(n) = \frac{1}{np} \sum_{j=1}^{p} \sum_{i=1}^{n} x_{ij} \qquad (15\text{-}3)$$

$$s^2(n) = \frac{1}{p-1} \sum_{j=1}^{p} [x_j(n) - m(n)]^2 \qquad (15\text{-}4)$$

The two estimates can then be used to establish a confidence interval.

Figure 15-3 shows the result of applying the procedure for the single server system. Results are shown for utilizations of 0.2, 0.3, 0.4, 0.5, and 0.6. In each

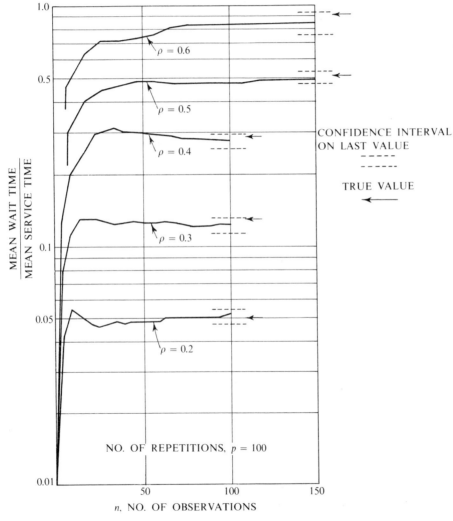

Figure 15-3. Accumulated mean wait time for system, starting from idle state.

case, the experiment has been repeated from an initial idle state, different random numbers being used on each repetition. The results show the estimated mean waiting time calculated from Eq. (15-3) as a function of sample size, n. Measurements were made in steps of $n = 5$ for $\rho = 0.2, 0.3$, and 0.4, and $n = 10$ for $\rho = 0.5$ and 0.6. Each case is for 100 repetitions ($p = 100$). Also shown are the 90% confidence intervals calculated for the highest value of n used in each case, and the known *true* values of the mean waiting times. It will be noticed that the true values in all cases do, in fact, fall within the confidence intervals.

The mean upon which the confidence interval is based depends upon $1/np$. In the absence of initial bias, the same proportionate increase in either n or p can be expected to have equivalent effects upon the size of the confidence interval. They can also be expected to cost approximately the same amount of computer time for their execution. However, to increase the probability of reducing the initial bias to the point where it can be considered negligible, it is preferable to extend the runs, holding the number of repetitions to a level at which the sample size is large enough to justify the approximation to a normal distribution. For a given amount of computing, the run size can then be maximized, reducing the effect of initial bias to a minimum. The criteria for making such choices are discussed more fully in Refs. (5) and (6).

15-5

Elimination of Initial Bias

The results of Fig. 15-3 clearly show the effects of initial bias. The fact that the true values fall within the estimated confidence intervals suggests that the runs were long enough to have made the initial bias negligible. The theoretical results used to establish this fact, however, are not usually available. Two general approaches can be taken to reduce the effect of initial bias. The system can be started in a more representative initial condition or the first part of each simulation run can be ignored.

In some simulation studies, particularly of existing systems, there may be information available on the expected conditions that make it feasible to select better initial conditions than the idle state. However, a range of values for the initial conditions should be used from which to choose a different initial state for each repetition. To use the same initial condition for each run may reduce the bias by removing an unusual starting condition but it will leave some degree of correlation between the runs. Unfortunately, this approach involves knowing a great deal about the system performance before starting the simulation. There are, however, cases where this approach can be used. In addition, it is possible to use the approach to check the accuracy of a simulation by reiteration. Having derived results by one method or another, the conditions predicted by the results can be used to indicate reasonable starting conditions. If the original results are truly independent of initial bias, the repetition of some of the runs with the new initial conditions should produce no significant difference.

The more common approach to removing initial bias is to eliminate an initial section of the run. The run is started from an idle state and stopped after a certain period of time. The entities existing in the system at that time are left as they are. The run is then restarted with statistics being gathered from the point of restart. As a practical matter, it is usual to program the simulation so that statistics are gathered from the beginning, and simply wipe out the statistics gathered up to the point of restart. No simple rules can be given to decide how long an interval should be eliminated. It is advisable to use some pilot runs starting from the idle state to judge how long the initial bias remains. This can be done by plotting the measured statistic against run length as has been done in Fig. 15-2. It is highly desirable, however, that the pilot investigation be done by repeating runs as has been done in Fig. 15-3. An examination of the three individual runs of Fig. 15-1 will show how difficult it is to judge from a single run when the measured value has approached its steady value. At the expense of a little more calculation, the presence of initial bias can be examined by studying the behavior of the standard deviation. In the absence of initial bias, the standard deviation can be expected to be inversely proportional to $n^{1/2}$. By examining the way the standard deviation changes with sample length, it is possible to see whether this relationship is being met. For example, Fig. 15-4 shows the standard deviation obtained at the time the results of Fig. 15-3 were derived. The logarithm of the standard deviation, derived from Eq. (15-4), is plotted against the logarithm of n. The result should approximate a straight line sloping downwards at the rate of 1 in 2 (for equi-scaled axes). It can be seen that the curves initially increase but eventually decline as expected.

Judged from the results of Fig. 15-4, reasonable allowances for an initial period would appear to be:

Utilization	Cut-off Point
0.2	7
0.3	14
0.4	30
0.5	60
0.6	120

The chosen values are probably conservative because they are accumulated statistics. Using the above values for an initial cut-off period, Figs. 15-5 and 15-6 show the results of repeating the experiments of Figs. 15-3 and 15-4.

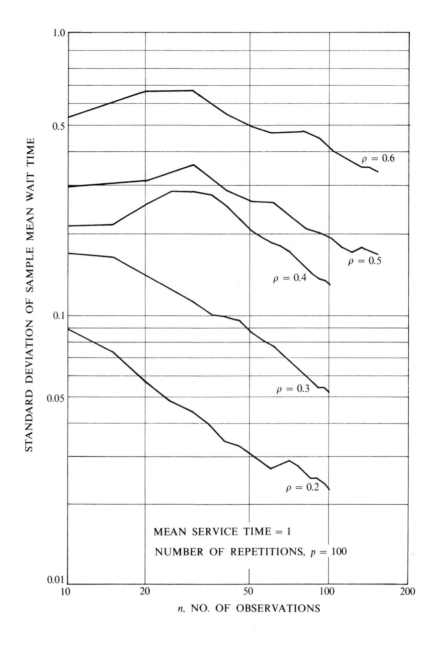

Figure 15-4. Variation of standard deviation with sample size, starting from idle state.

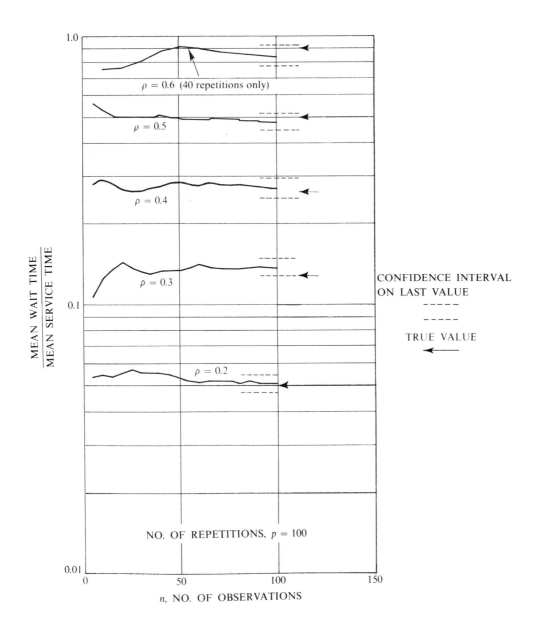

Figure 15-5. Accumulated mean wait time for system, with initial period removed.

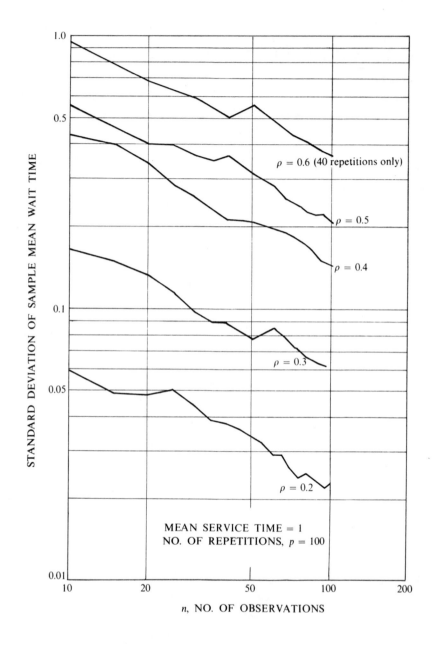

Figure 15-6. Variation of standard deviation with sample size, initial period removed.

Another approach to the problem of estimating the precision of simulation results does not rely upon repetition, but uses a single long run, preferably with the initial bias removed. The run is divided into a number of segments to separate the measurements into batches of equal size. The mean of each batch is taken and the individual batch means are regarded as independent observations. The estimated value of the variable being measured is the mean of the batch means. This, of course, is exactly the same as the mean of all the measurements. However, the assumption that the batch means are independent, together with the application of the central limit theorem, allows the batch mean observations to be treated as normally distributed. The usual formula can then be applied to estimate the variance of the mean and so compute a confidence interval.

The method cannot be applied to an accumulated statistic, such as the accumulated mean waiting time discussed in the previous section, because the distribution of the sample mean depends upon the run length and successive batch means cannot be treated as observations from the same population. Typically, the method might be used in measuring a queue length where the experiment is conducted by sampling the queue length at uniform intervals of time, so that each observation is an individual measure of the same random variable.

A complete run consists of N observations which are broken into p batches of size n, so that $N = np$. (It is assumed that N is exactly divisible by p.) In effect, the experiment is equivalent to repeating an experiment of length n a total of p times, with the final state of one run becoming the initial state of the next run. This way of repeating a run is preferable to starting each run from an initial idle state, since the state at the end of a batch is a more reasonable initial state than the idle state. However, the connection between the batches introduces correlation. Sometimes, the batches are separated by intervals in which measurements are discarded in order to eliminate the correlation. Clearly, this throws away useful information. Conway (1) demonstrates that the variance to be expected by using all the data and accepting the correlation between batches is less than that obtained from the reduced amount of data obtained by separating the batches. It seems to be preferable, therefore, to work with adjoining batches.

The batch mean method has the advantage of the repetition method without the necessity of eliminating the initial bias on each repetition. However, it is necessary to assume that the individual batch means are independent. The assumption can be justified if the batch length is sufficiently long. The effect of autocorrelation is that the value of one piece of data affects the value of following data. The effect diminishes as the separation between the data increases and beyond some interval size it may reasonably be ignored. If the batch size is greater than this interval, the batch means may be treated as independent. It remains a matter of judgement to choose a suitable batch size. It might reasonably be speculated that the interval over which the batch is measured should be at least as great as the interval excluded from the beginning of a run to remove initial bias. If that value has been determined, it can also be used as a batch size. However, the only safe procedure is to use a test run in which to try a batch size and test for the presence of correlation in the results (7). Another approach is to repeat the calculations with several batch sizes and test for consistency of results (8). By making the batch sizes multiples of each other, it is possible to perform the operation in a single run.

In discussing the repetition method, it was pointed out that there is a trade-off between the number of repetitions and the run length. With the batch method there is a similar trade-off between batch size and number of batches. Since the number of batches corresponds to the number of samples of an assumed normal distribution, it is again advisable to hold this number to a reasonable limit to meet the assumption and maximize the batch size, in order to reduce the correlation between batches.

An important practical aspect of the batch method is that it does not entail the simultaneous presence of all the data to carry out the calculations. The batch means can be calculated as the simulation run proceeds. Computer space is only required to accumulate the sum of the batch means and the sum of their squares, together with an accumulation of the numbers forming the current batch mean. If multiple batch sizes are collected, a set of three such numbers is needed for each batch size.

15-7

Time Series Analysis

A more recently developed approach to estimating the precision of simulation results is to estimate the variance of a sample mean, including autocorrelation effects, from results derived in the study of time series. The simulation experiment is conducted as a single run with the initial bias removed. However, the individual observations are preserved and treated as the data of a time series. For convenience, suppose the observations are made at unit time intervals and the record is for a time length T, so that there are T observations.

The autocorrelation is measured by a series of autocovariance coefficients which show the extent to which values separated by an interval of τ time units affect each other. The coefficients are defined by the formula:

$$R(\tau) = \frac{1}{T - \tau} \sum_{t=1}^{T-\tau} (X_t - \bar{X})(X_{t+\tau} - \bar{X}) \qquad (\tau = 0, 1, 2, \ldots, T - 1)$$

where X_t is the observation at time t and

$$\bar{X} = \frac{1}{T} \sum_{t=1}^{T} X_t$$

The particular case of $\tau = 0$, $R(0)$, is an estimate of the variance σ^2 of the distribution from which X_t is drawn.

The estimation of the sample mean variance from the autocovariance coefficients involves some computational subtlety, but an acceptable formula is (9), (10):

$$V(\bar{X}) = \frac{1}{T} \left\{ R(0) + 2 \sum_{\tau=1}^{M} \left(1 - \frac{\tau}{T} \right) R(\tau) \right\} \qquad (15\text{-}5)$$

The first term corresponds to the variance of the population from which X_t is drawn, and the term $R(0)/T$ estimates the variance of the sample mean that would be expected if the observations were independent. The additional term represents the contribution of the autocorrelation.

It will be noticed that Eq. (15-5) only includes the first M autocovariance coefficients (excluding $R(0)$). Normally the values of the coefficients decrease as τ increases. The value of M must be large enough to include the coefficients that are significant but, because of the finite size of the simulation run, M cannot be indefinitely increased without loss of accuracy. The choice of an appropriate value of M involves a decision similar to that involved in choosing a suitable batch size. The approach taken by Fishman (9) is to carry out the computation for a number of different values of M. The product $TV(\bar{X})$ should be independent of M, and a suitable value of M can be found by detecting that the product does not vary significantly with M. A detailed application of this method is given in Ref. (9) in which one case studied is the problem discussed in Sec. 15-4 with $\rho = 0.8$. For this high utilization it was found that M must be raised as high as 200 to 300.

Unlike the method of repeated runs or batch means, the analysis of autocorrelation requires that a considerable amount of data be preserved and analyzed after the simulation run.

15-8
Spectral Analysis

The analysis of time series through autocorrelation is intimately related to another way of viewing time series. A time series may also be regarded as the summation of oscillations of different frequencies. The spectrum of frequencies and the amplitudes of the oscillations can be formally related to the autocorrelation (10). Essentially the same calculations involved in the estimation of autocorrelation can be applied to derive a spectral analysis of a simulation run. The results can be used to test the precision of a measure by computing sample mean variances (11).

A spectral analysis, however, can provide more information than is contained in the estimate of a mean value. Comparing two systems on the basis of mean values of such factors as waiting time or queue length is a rather gross comparison. Two systems may show no significant difference in their mean values, but their transient behavior may be significantly different. One system may respond slowly to deviations from its mean values while the other may respond rapidly. Depending upon the purpose of the system, one performance may be preferable to the other. A simple comparison of means will mask the difference but a spectral analysis can distinguish the difference by showing whether the spectrum emphasizes low or high frequencies.

The technique of performing spectral analyses on simulation runs and the testing of significant differences is discussed in Ref. (11). As in the case of computing autocorrelation, there are several problems in the application of numerical computation methods, and a considerable amount of data must be preserved for analysis. The insight obtained in the behavior of a system, however, is very significant.

Bibliography

1 Conway, R. W., "Some Tactical Problems in Digital Simulation," *Management Science*, X, No. 1 (Oct., 1963), 47–61.

2 Burdick, Donald S., and Thomas H. Naylor, "Design of Computer Simulation Experiments for Industrial Systems," *Comm. ACM*, IX, No. 5 (May, 1966), 329–338.

3 Naylor, Thomas H., Kenneth Wertz, and Thomas H. Wonnacott, "Methods for Analyzing Data from Computer Simulation Experiments," *Comm. ACM*, X, No. 11 (Nov., 1967), 703–710.

4 Diananda, P. H., "Some Probability Limit Theorems With Statistical Applications," *Proc. Camb. Phil. Soc.*, XLIX (1953), 239–246.

5 Geisler, Murray A., "The Sizes of Simulation Samples Required to Compute Certain Inventory Characteristics with Stated Precision and Confidence," *Management Science*, X, No. 2 (Jan., 1964), 261–286.

6 Fishman, George S., "Digital Simulation: Input–Output Analysis," Tech. Memo. RM 5540PR, Santa Monica, Calif.: Rand Corp., Feb., 1968.

7 Conway, R. W., B. M. Johnson, and W. L. Maxwell, "Some Problems of Digital Machine Simulation," *Management Science*, VI, No. 1 (Oct., 1959).

8 Mechanic, H., and W. McKay, "Confidence Intervals for Averages of Dependent Data in Simulations—II," Report No. ASDD 17-202, IBM Corp., Yorktown Heights, N.Y., 1966.

9 Fishman, G. S., "Problems in the Statistical Analysis of Simulation Experiments: The Comparison of Means and the Length of Sample Records," *Comm. ACM*, X, No. 2 (Feb., 1967), 94–99.

10 Jenkins, G. M., "General Considerations in the Analysis of Spectra," *Technometrics*, III, No. 2 (May, 1961), 133–166.

11 Fishman, George S., and Philip J. Kiviat, "The Analysis of Simulation-Generated Time Series," *Management Science*, XIII, No. 7 (March, 1967), 525–557.

INDEX

G